THE CLIMATE ACTION HANDBOOK

THE CLIMATE ACTION HANDBOOK

A VISUAL GUIDE TO 100 CLIMATE SOLUTIONS FOR EVERYONE

Heidi A. Roop, PhD

ILLUSTRATED BY Joshua M. Powell

SASQUATCH BOOKS | SEATTLE

For Skip, whose quest for climate knowledge is insatiable and whose unconditional love and support has propelled me through my climate career.

And for Abbie, whose future I'm fighting for.

Printed in the United States of America

SASQUATCH BOOKS with colophon is a registered trademark of Penguin Random House LLC

27 26 25 24 23 9 8 7 6 5 4 3 2 1

Illustrator: Joshua M. Powell
Editor: Jen Worick
Production editor: Peggy Gannon
Designer: Tony Ong

Library of Congress Cataloging-in-Publication Data

Names: Roop, Heidi A., author. | Powell, Joshua M., illustrator.
Title: The climate action handbook : a visual guide to 100 climate solutions for everyone / Heidi A. Roop, Phd., illustrated by Joshua M. Powell.
Description: Seattle : Sasquatch Books, [2023] | Includes index.
Identifiers: LCCN 2022022741 (print) | LCCN 2022022742 (ebook) | ISBN 9781632174147 (paperback)
Subjects: LCSH: Environmental protection--Citizen participation. | Climate change mitigation. | Climatic changes. | Environmental responsibility.
Classification: LCC TD171.7 .H435 2023 (print) | LCC TD171.7 (ebook) | DDC 363.7/0525--dc23/eng/20220812
LC record available at https://lccn.loc.gov/2022022741
LC ebook record available at https://lccn.loc.gov/2022022742

ISBN: 978-1-63217-414-7

Sasquatch Books
1325 Fourth Avenue, Suite 1025
Seattle, WA 98101
SasquatchBooks.com

FSC
www.fsc.org
MIX
Paper from responsible sources
FSC® C005010

CONTENTS

Energy Production and Transportation 13

Travel and Work 33

Food and Farming 61

Civic and Community Engagement 179

Education and Climate Information 195

PREFACE

IN 2015, 196 countries adopted the Paris Agreement, committing to the goal of limiting the increase in global temperature to well below 2°C (3.6°F) and "to pursue efforts to limit the temperature increase to 1.5°C" (2.7°F) above pre-industrial levels of the late 1800s. If the rate of current warming continues, global average temperature is likely to reach the 1.5°C (2.7°F) target between 2030 and 2052—within the lifetime of most people alive today. With every increment of additional warming, the costs and challenges of climate change increase. The next decade is pivotal for climate action and for global society. What we choose to do—or not do—over the next several years will shape our collective climate-changed future for generations. You may ask, what can I do to help? This book aims to help you answer that question.

Meeting our global climate goals can be achieved only if action is taken to drastically reduce emissions of greenhouse gases like carbon dioxide (CO_2) and methane (CH_4). The Intergovernmental Panel on Climate Change (IPCC), the United Nations body for evaluating the science related to climate change, notes that carbon dioxide emissions need to be reduced by around 45 percent from 2010 levels by 2030, and to net zero by 2050 in order to limit warming to 1.5°C. Net zero means that the amount of carbon dioxide being produced is matched by the amount being removed, by activities that take it out of the atmosphere, like planting trees, and deploying technologies that remove CO_2 from the atmosphere.

This all requires societal transformation and rapid implementation of ambitious greenhouse gas reduction measures. Critically, no single sector, action, or fuel type can provide the needed emission reductions to reach these targets. Limiting warming to 1.5°C will require largely phasing out coal use by 2050, reducing CO_2 emissions from industry by 75 to 90 percent by 2050 (relative to 2010), supplying most electricity from renewable energy sources like wind and solar, and significantly enhancing energy efficiency across all sectors.

Much of this work is underway, but reaching these goals means we must accelerate the pace and scale of our global climate work over the next 10 to 20 years. While there is no question that this type of transformative change will require large-scale systems-based changes, each and every one of us can show up in the conversations and actions needed to help meet these ambitious goals.

Something you care about is at risk from climate change. We each need to learn about the local and global consequences of these changes and find ways, large and small, to engage in solutions. This book is an attempt to outline just a few of the myriad ways that individuals from all walks of life and perspectives can better understand what climate change will mean for the things they care about, as well as to map out ways we contribute to climate solutions. This book represents a passion project meeting a practical need. I, too, wanted to know where there were opportunities for me to learn, engage, and act. I wanted to help answer the question I hear daily: What can I do to help address climate change?

The climate actions presented here attempt to paint a picture of the breadth of opportunity that abounds for climate action. Importantly, not all climate actions captured here are directly associated with emissions reductions; climate solutions aren't only an emissions-related numbers game, and not all actions are easily quantifiable. This book intentionally doesn't use a hierarchy or quantification of what actions are "best" or have the biggest "impact." Other resources, like Project Drawdown, an award-winning climate nonprofit, paint a sophisticated, data-driven picture of climate solutions at scale. Some of those same solutions are here in this book, but so are actions like being a savvy consumer of climate information, becoming civically engaged, and participating in climate conversations. Some actions are costly; some are free. Some are time-intensive; some require a little forethought or reflection. All have value and represent a form of impact whether environmental, social, individual, or collective.

Ultimately, the sections in this book are bite-size pieces to help us get our arms around a big, beautifully complex challenge. For each fact or statistic in this book, there were dozens more that couldn't fit. Each action or idea offers an entry point into a topic that might be of particular interest to you. Some might inspire you to think (and act!) in a different way. Others might just provide you with more knowledge around the impacts and consequences of climate change. There is a wealth of information available and numerous nuances for each action presented here. Each could easily have become a book in its own right.

A good news climate story is that new data, climate commitments, and solutions are emerging almost daily. Nothing in the climate space is static. Even over the time it took to write this book, new research, poll results, and climate commitments were made. This will continue. I choose to see this as a sign that the climate change landscape is shifting and evolving to meet the challenge of the moment, with new interest and investments being made to confront this grand challenge. We are living through a dynamic moment that will surely make it into history books, but we still get to determine what will be written.

Importantly, facts and figures don't necessarily change hearts and minds, but people doing things can. Many of the people I interact with, from all different walks of life, are seeking more information to inform their unique climate solutions journey. They are often looking for more information about specific actions to inform themselves, but also to share in conversations with their family members, friends, and coworkers, and in their communities.

Let's be clear, climate change is bringing environmental, social, and cultural change. Past emissions of greenhouse gases have committed us to change, but we still get to choose how much change we experience, and what kind. Human-caused climate change is already causing loss of livelihoods, culture and identity, life, and property. But embedded within many climate solutions are opportunities—to create a healthier, brighter future. Climate actions can create near-instantaneous improvements in air quality and can offer ways to address societal inequities, green our communities, save money, build local economies, and develop or deepen our personal relationships.

My hope is that something in the pages that follow will empower you to evaluate, engage, and act. How we engage and what we do may change day-to-day, month-to-month, or year-to-year, but what matters is that we act and that we build and maintain momentum. We can do that by each doing some of the work as individuals and working with others to advocate for the systems-based changes we know are needed to decarbonize the world and build climate-ready communities. Together, we can help shape and create the future we want through actions big and small. We just have to lean in, listen, learn, and act. What the future holds is not yet determined. When it comes to climate change, we still choose what the future holds. How will you help shape the future?

OVERVIEW

GLOBAL GREENHOUSE GAS EMISSIONS BY SECTOR

IN 2018

18.4% AGRICULTURE, FORESTRY and LAND USE

LIVESTOCK & MANURE • SOILS • RICE CULTIVATION • CROP BURNING • DEFORESTATION • CROPLAND • GRASSLAND

3.2% WASTE

LANDFILLS • WASTEWATER

5.2% INDUSTRY

CHEMICALS • CEMENT

15.3% OTHER ENERGY USE

UNALLOCATED FUEL COMBUSTION • FUGITIVE EMISSIONS FROM ENERGY PRODUCTION • ENERGY IN AGRICULTURE & FISHING

17.5% ENERGY USE IN BUILDINGS

RESIDENTIAL BUILDINGS • COMMERCIAL BUILDINGS

16.2% ENERGY USE IN TRANSPORT

ROAD TRANSPORT • AVIATION • SHIPPING • RAIL • PIPELINE

24.2% ENERGY USE IN INDUSTRY

IRON & STEEL • NON-FERROUS METALS • CHEMICAL & PETROCHEMICAL • FOOD & TOBACCO • PAPER & PULP • MACHINERY

GREENHOUSE GASES: WHY ARE THEY SUCH A PROBLEM?

UNDERSTANDING HUMAN-CAUSED climate change requires understanding the root cause of the problem: heat-trapping greenhouse gases, like carbon dioxide, methane, nitrous oxide, and water vapor. All of these heat-trapping gases occur naturally, but the problem today is the rapid increase in their production since the Industrial Revolution. This is due to the combustion of fossil fuels—like coal, oil, and natural gas—for transportation, electricity generation, food production, and commercial, residential, and industrial activities, and from changing land use patterns, including deforestation.

Why do we care about heat-trapping gases? There is a clear relationship between the amount of greenhouse gases emissions and the warming of the planet. For the last 130 or so years, humans have been pumping heat-trapping gases into the atmosphere in quantities that have pushed the concentration of greenhouse gases in the atmosphere higher than has been recorded in geologic records spanning back millions of years. Collectively, we emit about 50 billion metric tons of greenhouse gases every year, often reported in carbon dioxide equivalent (CO_2e), which provide a single measure for converting the warming impact of different greenhouse gases. This represents a 40 percent increase in emissions since 1990. This is not good for people and the planet.

Humans are producing massive quantities of these heat-trapping gases, and we can already see the oceans, atmosphere, and land warming as a result. By 2021, global average temperature had increased nearly 1.1°C (2°F) since 1850. We expect this warming to continue until we drastically reduce emissions of greenhouse gases and actively work to remove them from the atmosphere. Greenhouse gases like carbon dioxide remain in the atmosphere for hundreds of years, so once we emit, we commit. We commit to further warming, but also to the climate impacts that stem from this warming, like intensifying hurricanes, floods, droughts, and heat waves.

We know the source of the problem. We know the sectors from which most of these gases are produced, like transportation, electricity generation, industry, and agriculture. We know that just 100 companies are responsible for 71 percent of total global industrial emissions of greenhouse gases since 1988. Critically, we know that we need to deploy solutions that reduce emissions, but we also need to prepare for the climate changes already set in motion. We know that the problem is urgent, and that the sooner we act, the easier it will be to avoid the worst impacts of climate change. Simply put, the impacts of a warming world are everywhere, and we know the cause of this warming: us. The time to act is now. And that's on us too.

UNDERSTANDING THE SCALE OF THE PROBLEM

A 2021 report by the Intergovernmental Panel on Climate Change (IPCC) states, "It is unequivocal that human influence has warmed the atmosphere, ocean, and land." This report—one in a series of reports related to the state of our knowledge about climate change, which are updated every six to seven years—was written by 234 authors from 65 countries. These authors, experts from a range of disciplines, volunteer their time and expertise to the IPCC. For this report, the authors synthesized the findings of 14,000 scientific papers!

Thanks to their hard work, we can boil down the current state of the science into the following:

- **IT'S US.** It is unequivocal that the burning of fossil fuels has warmed the planet.
- **IT'S HERE.** There are clear connections between our warming world and changing patterns in extremes like heat waves, heavy precipitation, and droughts.
- **IT'S EVERYWHERE.** Human-caused climate change is already affecting weather and climate extremes in every region of the world.
- **WE'VE COMMITTED TO CHANGE.** Global surface temperature will continue to increase until at least the midcentury under all emissions scenarios. We've locked in warming because CO_2 is a long-lived greenhouse gas, meaning it stays in the atmosphere for a long time. Unless we physically remove it from the atmosphere, it will continue to trap heat and warm the planet.

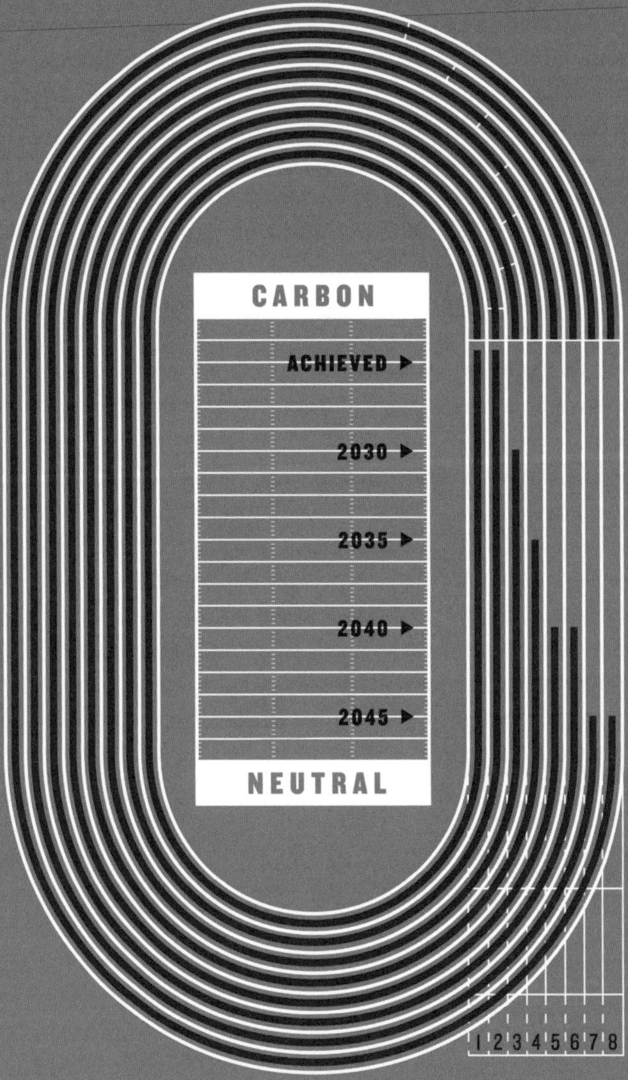

AS OF 2021, EIGHT COUNTRIES HAVE ACHIEVED NET ZERO OR PLEDGE TO DO SO BEFORE 2050

1. SURINAME 2. BHUTAN 3. URUGUAY 4. FINLAND 5. AUSTRIA 6. ICELAND 7. GERMANY 8. SWEDEN

THE UNITED STATES AND 123 OTHER COUNTRIES HAVE PLEDGED TO ACHIEVE NET ZERO BY 2050

- **THE MORE WE EMIT, THE WORSE IT GETS.** Just as we know that we've committed to a certain amount of warming, we know that how much warmer the world will get is directly related to the amount of greenhouse gas emissions we create as a global society.
- **WE STILL CHOOSE, BUT THERE'S NO TIME TO WASTE.** Because we clearly know the source of the problem, we can also deploy solutions. The opportunity we have over the next decade is that we get to determine how much future warming is "locked in" beyond 2050. Urgent action is needed. We need to reduce global emissions of greenhouse gases by about half by 2030 to avoid the worst consequences of a warming world.

While the scale of the problem can be used to rationalize inaction or a sense of defeat, if we have any hope of creating a safe, healthy future for ourselves and for future generations, we know we need to act—as individuals, as communities, as counties, townships, and boroughs, as states, as countries, and as global residents. In an intimately interconnected globalized world, the costs and consequences of faraway climate impacts can find their way into our lives. Climate change needs no passport. We should all be motivated to act.

UNDERSTANDING THE INEQUITIES OF CLIMATE CHANGE

Climate change is touching every region of the globe, including our very own communities. As we seek to prevent further warming and prepare for various climate impacts, we must also understand that who is most at risk to these various climate change-related impacts is deeply inequitable. Historical inequities, systemic racism, colonialism, and divestment mean that historically marginalized communities—including those people whose incomes are below the federal poverty threshold, Indigenous people, and people of color—are often already experiencing the negative impacts of climate change and face greater risks than other communities and individuals. These groups also tend to have the fewest resources to prepare and prevent further harm, and to recover when disaster strikes. These are the disproportionate impacts of climate change.

Simply put, where you live and work, the color of your skin, and the community or country you come from influence the risks you face from climate change. There is increasing recognition of these environmental injustices and inequities. Seventy percent of Americans support "increasing federal funding to low-income communities and communities of color who are disproportionately harmed by air and water pollution."

The climate and environmental justice movements require our attention, support, continued engagement, and investment. Climate justice, which requires removing systemic barriers and inequities, needs to be at the heart of our climate work. Environmental justice expert Jalonne White-Newsome outlined a powerful framework for us all to continually consider in a process she calls "ADAPT-ing." In a 2021 Carbon Brief article on climate justice, she said: "I truly believe that each of us—in whatever role we play—has the power to achieve climate and environmental justice by ADAPT-ing:

Acknowledging the harm
Demanding accountability
Addressing racism, power, and privilege
Prioritizing equity
Transforming systems"

We all need to center and prioritize "ADAPT-ing" in our climate work.

CLIMATE ACTION IN FOCUS: CLIMATE CHANGE MITIGATION AND ADAPTATION

Climate change science and solutions are full of jargon and often hard-to-understand words. However, climate actions broadly fit into two categories: climate change mitigation and climate adaptation. Put more simply, we need to consider actions that prevent the problem from getting worse (climate mitigation) while also

1. ALASKA

Warming faster than any state in the US, climate change in Alaska means melting glaciers and permafrost, thinning sea ice, ocean acidification, larger and more frequent wildfires, and loss of critical habitat for animals like bears and moose. Climate change is expected to cost Alaska billions of dollars and negatively impact human health, subsistence activities, critical infrastructure, culture, and threaten the state's food and water security.

2. NORTHWEST

Warming nearly 2°F since 1900, climate change means warmer winters, less snowpack, changing wildfire risk, warmer ocean and stream water, and rising sea levels. These changes threaten iconic species like salmon, stress water resources, and harm human health, cultural heritage, agriculture productivity, and outdoor recreation. Further warming is expected to melt the region's glaciers, increase flood, wildfire, and drought risk, and accelerate sea level rise.

3. NORTHERN GREAT PLAINS

From row crops to mountaintops, water is a critical part of the climate change story in the northern Great Plains. Across this region, flood and drought risk are increasing, the region's glaciers are melting, and changing seasons are both benefiting and challenging the agricultural economy. Warming is negatively impacting livelihoods and cultural heritage, reducing streamflow, stressing trout and other aquatic species, threatening critical energy infrastructure, and harming the region's rich recreational opportunities like hunting, fishing, and skiing.

10. HAWAII & US-AFFILIATED PACIFIC ISLANDS

Increased risk of extreme drought and flooding, sea level rise, warming oceans, and changing rainfall patterns pose a threat to human health, food and water security, local economies, and cultural heritage. Several unique species and the region's land and marine biodiversity are under threat, with some facing population decline or extinction. By 2100, 1 to 4 feet of sea level rise is expected, with consequences for livelihoods and infrastructure,increasing the risk of displacement, relocation, and migration.

9.SOUTHWEST

Representing one-fifth of the land area in the US, the Southwest spans the Pacific Coast, Sierra Nevada, and Sonoran Desert. Climate change is bringing warmer temperatures, rising seas, increased risk of droughts, heatwaves and wildfires, and declining snowpack. This threatens human health, cultural traditions, and the region's vital natural resources, including water, which supports the region's 60 million residents and a significant portion of the nation's fruit, nut, and vegetable production.

CLIMATE IMPACTS ACROSS

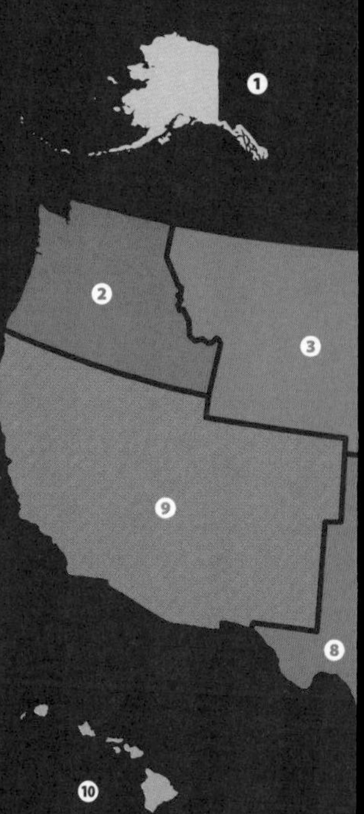

4. MIDWEST

The Midwest region is getting warmer, wetter, and more humid. The region faces more heavy rain events, warming lakes and rivers, increased pests, and stresses on humans and wildlife. Extreme heat and poor air quality threaten the health and well-being of many of the region's residents. A warmer future will further threaten the region's cultural heritage and human health, agricultural productivity, forests, grasslands, and wildlife, and will strain critical infrastructure like roads and stormwater systems.

5. NORTHEAST

The Northeast is both the most densely populated and forested in the US. From skyscrapers and shorelines to snow-covered mountains, climate change—particularly the shrinking contrast between seasons—is harming the region's culture, ecosystems, recreation, local economies, and natural resources. The Northeast is experiencing some of the highest rates of ocean warming and sea level rise in the United States. Further warming will worsen impacts on human health, critical infrastructure, and the region's economy.

6. SOUTHEAST

Across beaches, bayous and cities, climate change in the Southeast stresses human health, the built environment, forests, coasts, and the region's rich cultural heritage. More extreme heat, changing sea levels, heavier precipitation, and elevated flood risk are just a few signals of the Southeast's changing climate. Expected to suffer some of the largest economic losses from climate change in the United States, residents, industries, and infrastructure will all face challenges from a warming world.

THE UNITED STATES

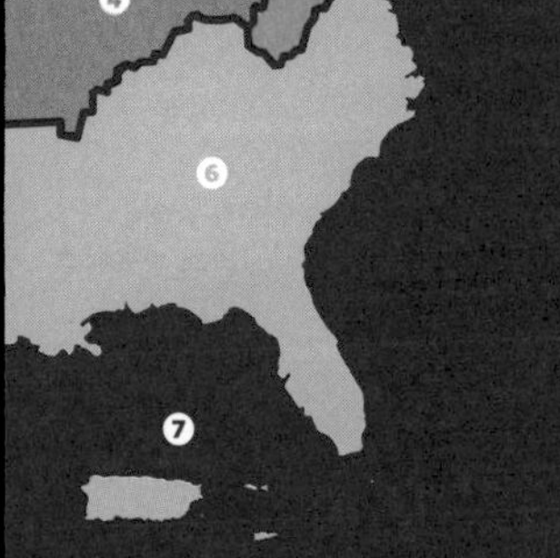

8. SOUTHERN GREAT PLAINS

A region that experiences hurricanes, floods, blizzards, droughts, and heatwaves, the southern Great Plains is facing many climate change-related risks: increasing temperatures and dryness, increased frequency of extreme heat, stronger hurricanes, and rising sea levels. This region is responsible for 25 percent of all US energy production and is a vital corridor for road and rail freight. Climate change is expected to impact infrastructure and critical services, local economies, and the health and well-being of its residents.

7. CARIBBEAN

With several large islands and over 800 small islands and cays, the US Caribbean is rich with marine and land-based biodiversity, unique forest habitat, ecotourism, and cultural heritage. With expected drier conditions, accelerated sea level rise, increased hurricane intensity, more extreme heat, and continued warming of the ocean, fisheries and other natural resources, local economies, and the health of the region's 3.5 million inhabitants are at increasing risk of negative climate-related impacts, including food security and freshwater access.

embracing solutions that help us all prepare for the climate changes we know we've already set in motion (climate adaptation).

Mitigation involves actions that reduce and remove heat-trapping greenhouse gases from the atmosphere. Major sources of greenhouse gases include cars, trucks, and airplanes; generation of electricity from coal and natural gas; food production and livestock; industrial activities like cement and steel production; and land use changes. Preventing further planetary warming requires rapid reduction in emissions of these heat-trapping gases from across all of these sectors and sources.

Climate adaptation actions involve preparation and seek to reduce our exposure to the impacts of climate change. This can include preparing our built environment—including our homes, roads, bridges, and water systems—to withstand the increasing stresses of events like floods, heat waves, droughts, and wildfire. Adaptation is also needed to make our farmlands, forests, and lakes, along with our communities and economy, better able to withstand the increasing threats and challenges that a warmer world will bring. This can include planting and breeding new crops that are more drought- or heat-tolerant, planting new tree species that will be better adapted to warmer climate conditions in the future, or working to ensure we have strong social and community networks to help us "weather the storm." Adaptation solutions involve an exciting range of opportunities to design and build healthier, stronger, and more equitable communities.

Climate change mitigation and climate adaptation often overlap when an action or intervention has co-benefits or multiple benefits. For example, improving the health of soils can mitigate climate change by storing carbon, while also reducing flood risks by making soil better able to soak up excess rainfall, reducing potential flood-related damages. Healthy soils represent an opportunity for both climate mitigation and climate adaptation.

The time to act and scale up successful mitigation and adaptation actions is now. We still choose how bad it gets, but there's no time to waste. Read on to get inspired to help prevent, and prepare for, the impacts of climate change. Fortunately, climate solutions surround us.

WE CAN STILL CHOOSE
BUT THERE'S NO TIME TO WASTE

STARTING AND SUSTAINING YOUR CLIMATE ACTION JOURNEY

EVEN SMALL ACTIONS CAN HELP MOVE THE DIAL ON CLIMATE CHANGE

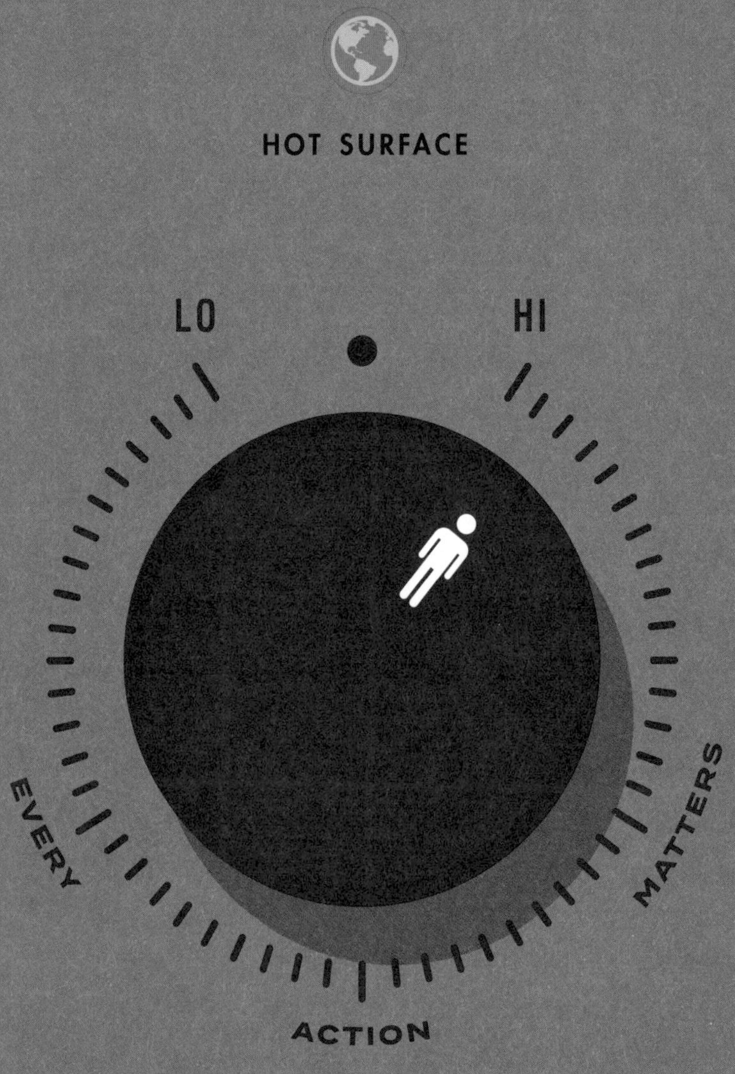

ACTION 1

CONSIDER COLLECTIVE AND INDIVIDUAL ACTIONS

This book was born out of a desire to explore the diversity of answers to these two common questions about climate change: What can I do to help address climate change, and do individual actions even matter? Large systems and structures—like our energy infrastructure, transportation systems, and corporate and governmental processes—need to fundamentally change to deploy solutions at the scale needed to confront the stark realities of climate change. Individuals, after all, are said to be "statistically blameless" for causing climate change, and individual actions alone will be insufficient to move the dial on climate change without broad governmental and regulatory interventions.

While there is warranted debate about the role of individuals and evidence that corporations benefit from the narrative that puts the burden of responsibility on individuals, I'm of the opinion that *every* action matters. Yes, some actions are small as far as their contribution to reducing emissions, but even seemingly small actions—like talking about climate change with our families and friends, or becoming more civically engaged—can help move the dial on climate change at home *and* at scale. Engagement and action can turn despair into determination. Determination and passion, paired with civic engagement and political action, can be a force to turn the tide on climate change.

This is especially true of action by those living in high-income, historically high-emitting countries. In fact, award-winning Project Drawdown notes that "individual and household actions have the potential to produce roughly 25–30 percent of the total emissions reductions needed to avoid dangerous climate change (>1.5°C rise)." While there is no question that our elected leaders and the corporations that produce the lion's share of emissions need to step up and stop using their influence and financial resources to stall or prevent progress on climate action, we as individuals can be agents of powerful change.

▸ **As you explore this handbook and choose what you *can* do, remember, *every* action matters.**

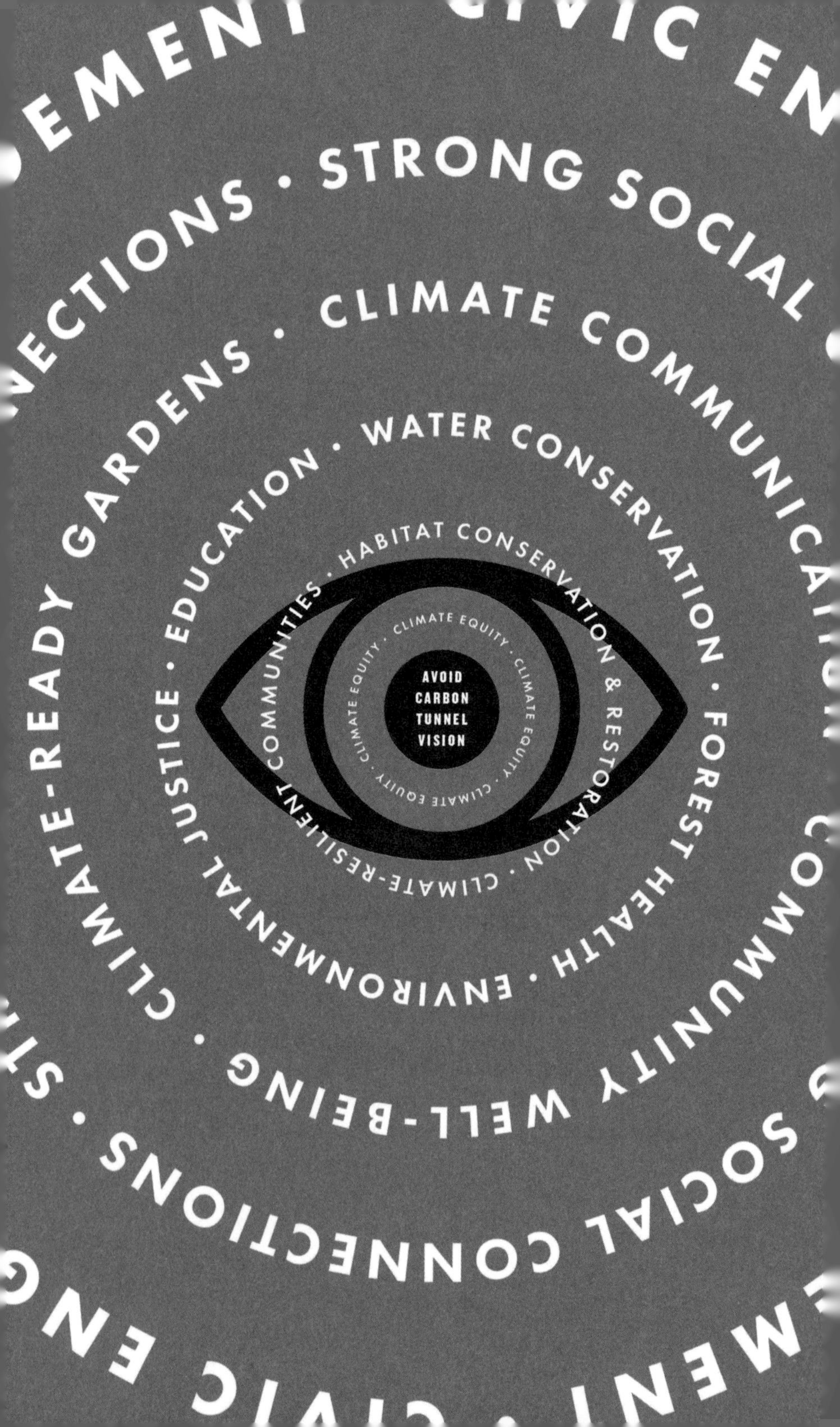
CIVIC ENGAGEMENT · STRONG SOCIAL CONNECTIONS ·
CLIMATE COMMUNICATION · COMMUNITY WELL-BEING · CLIMATE-READY GARDENS ·
WATER CONSERVATION · FOREST HEALTH · ENVIRONMENTAL JUSTICE · EDUCATION ·
HABITAT CONSERVATION & RESTORATION · CLIMATE-RESILIENT COMMUNITIES ·
CLIMATE EQUITY · CLIMATE EQUITY · CLIMATE EQUITY · CLIMATE EQUITY ·
AVOID CARBON TUNNEL VISION

ACTION 2

CENTER ACTION IN YOUR STRENGTHS AND PASSIONS

There are myriad climate solutions and climate actions, and each of us is equipped with different interests, skills, and resources that can make a difference. All this beautiful diversity of perspective, experience, and passion is needed to address this grand societal and environmental challenge. The hard part is making sure you find ways to engage and act that keep you contributing in ways big and small for the long haul.

Our actions should also leave us personally inspired and motivated. This can be easier said than done as the realities of climate change are grim and can take a real mental and emotional toll (see Action 76). Many climate actions can also feel sacrificial. While there is no question that giving up some things we love or scaling back high-emitting activities is needed, as with any grand challenge, the key to successful climate work is sustained action and engagement.

Climate actions, as highlighted throughout this handbook, can span everything from reducing air travel or retrofitting a home to engaging in climate storytelling or art and finding ways to deepen connections within your community. While it may seem easier to understand or quantify our contributions by just looking at the amount of emissions reduced, we want to avoid carbon tunnel vision. There are many other facets of climate work that are also meaningful contributions. Investing time, energy, and emotion in helping shape the future of your community—say through being engaged in a community climate action plan or through sharing your own creative talents or advocacy efforts—is critical in a balanced portfolio of climate solutions work.

Seeking space to engage in climate solutions can start with identifying your strengths, passions, and motivations. Then bring those strengths and passions to our collective table: climate change is an all-hands-on-deck situation, and what we shape and create together needs to include a rich diversity of perspectives and approaches. Critically, we don't want to stay siloed or stagnant in our work. Building your climate community can also provide the motivation to sustain you on your climate journey.

▸ **What strengths and passions will you bring to your climate work? Who is in your climate community?**

ACTION 3

UNDERSTAND THE DISCONNECT BETWEEN ACTIONS AND IMPACT

Individual climate action and systems-based changes are both needed to turn the tide on human-caused climate change, but there can be a real disconnect between our actions and our impact. Our knowledge and perceptions of what "high-impact" actions are, like those that result in the highest energy savings and reduction in emissions, can be incorrect.

For example, a 2010 survey of adults across the United States asked participants to estimate energy use for common household devices like lightbulbs and laptops, as well as energy savings for a subset of common household activities, like washing clothes in cold water and turning down the thermostat. The key finding was that many of us generally underestimate energy use and savings, but many of us also tend to focus on lower-impact "curtailment" activities, like turning off lights and driving less, compared to activities that increase energy efficiency, like installing more energy-efficient appliances which tend to result in comparatively higher emissions reductions. While it is true that all of these activities have some impact, this, and other research, indicates that we generally misunderstand the impact of different actions and their effectiveness as climate solutions.

To make matters worse, a 2017 study noted that across the European Union, the United States, Canada, and Australia, a majority of government resources on climate change focused primarily on solutions with the least impact on emissions reductions.

Beyond perusing the actions in this book and engaging in solutions that work for you, your family, and your community, remember that we need to engage in a range of actions and solutions, from emissions reductions to strengthening our connections with our communities. We also have to resist a one-and-done attitude, where engaging in a single action limits our sense of urgency and motivation to engage in multiple actions. This single-action bias can be a big barrier to engaging in a diversity of solutions, which are needed to both meaningfully reduce emissions and to prepare for climate change.

▸ **What actions will you engage in to avoid the pull of single-action bias?**

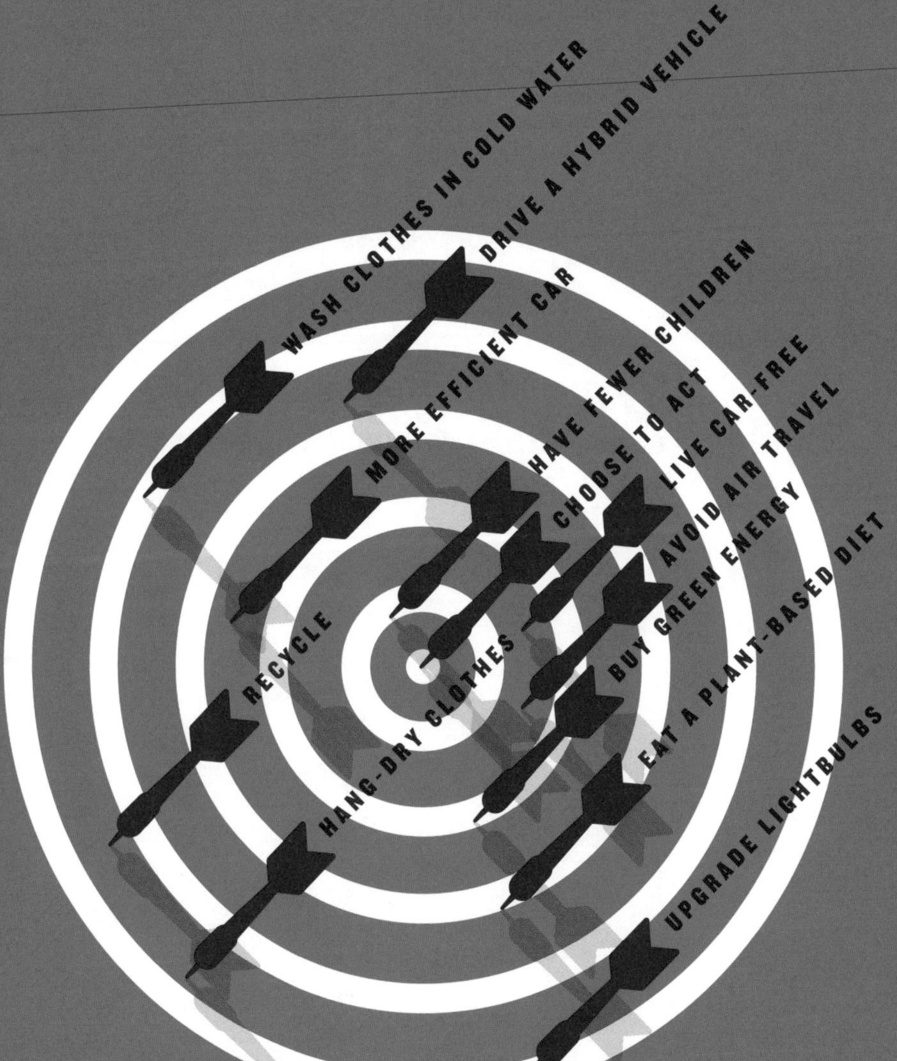

DIFFERENT ACTIONS REDUCE OR AVOID DIFFERENT AMOUNTS OF GREENHOUSE GAS EMISSIONS

70%

OF REGISTERED VOTERS SUPPORT TRANSITIONING THE US ECONOMY FROM FOSSIL FUELS TO CLEAN ENERGY BY 2050. YET, 90% OF REGISTERED VOTERS IN THE UNITED STATES HAVE NOT CONTACTED THEIR ELECTED OFFICIALS TO ACT. WILL YOU?

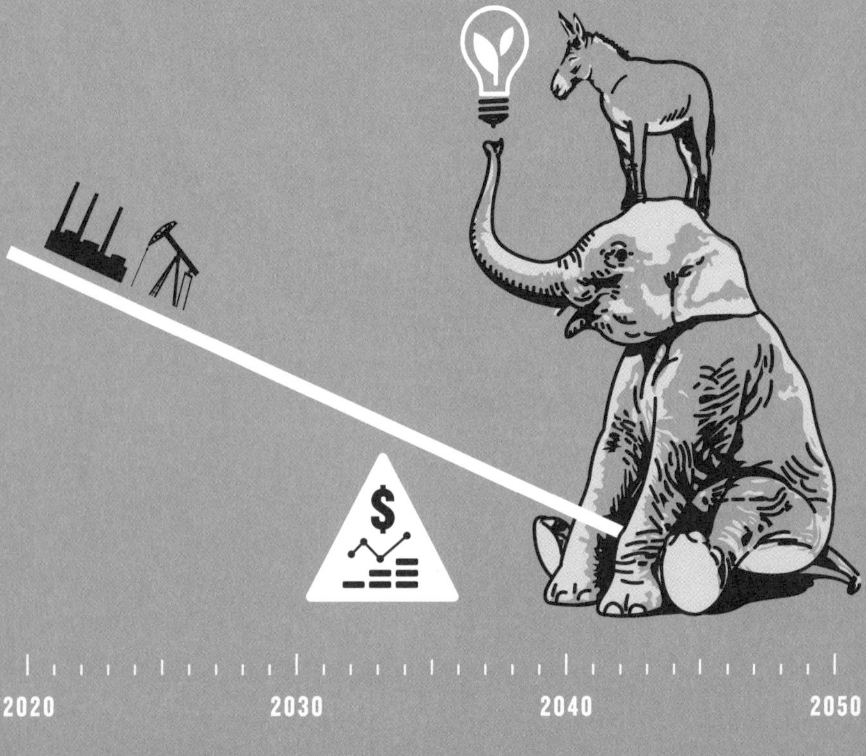

ACTION 4

BE PRIVY TO THE POLITICS OF CLIMATE CHANGE

In April 2021, President Biden announced a new greenhouse gas reduction target for the United States. This new goal sets out to achieve a 50 to 52 percent reduction in greenhouse gas emissions from US activities by 2030 (relative to 2005 levels), on top of commitments to achieve "net zero emissions economy-wide" by 2050 and "a carbon pollution–free power sector" by 2035. Such bold commitments and corresponding action and investments are required if we stand a chance of limiting global average warming well below 2°C (3.6°F), the current global policy target as agreed to under the Paris Agreement.

While the United States ranks first in the share of historical emissions (about 20 percent of total global emissions since 1850), US actions and leadership are critical to meeting these commitments. Investments like the Inflation Reduction Act that was signed in late 2022, which represents the single largest investment by the US government to reduce greenhouse gases, are a step in the right direction for helping meet these critical climate goals. Despite this historic US climate investment, the bill only narrowly passed through the US Congress with votes falling along party lines. There was unanimous Republican opposition to the bill.

While there remains a partisan divide when it comes to climate change, there is an increasing level of concern among constituents across the political spectrum about climate change and its impacts on people, the environment, and the economy. There is also a clear desire to see more climate action across all levels of government and by individuals, regardless of political affiliation. Recent polling of Americans across the political spectrum shows that 64 percent want the United States to do more, 61 percent want more local government action, 60 percent want to see more state governors take action, and 69 percent want to see individuals doing more. When asked to look in the proverbial mirror, 55 percent indicated that they, themselves, should do more.

Despite this bipartisan interest in more action, 90 percent of registered voters have not yet contacted their elected officials to urge them to act on climate.

▸ **Will you avoid the partisan pull around climate change, and help advocate for action by all?**

ACTION 5

BEWARE THE CORPORATE "ANTI-CLIMATE" CAMPAIGN

While climate change remains a partisan topic with US elected officials, we also have to contend with a highly funded, coordinated campaign against the science and scientists who have documented–using multiple lines of evidence–that the world is warming and that this warming is unequivocally associated with human activities.

Who is coordinating this anti-climate campaign? Recent exposés highlight that the fossil fuel industry and the organizations, think tanks, and public relations firms they bankroll actively spread misinformation and deceive consumers in order to prevent or delay a transition to cleaner energy. ExxonMobil is reported to have clearly known about, through their own internal research, the negative impacts of greenhouse gas emissions produced by their products since the 1970s. Today, fossil fuel companies and other high-emitting corporations have funded "climate-friendly" ads, "educational" resources, and tools like online carbon footprint calculators. These marketing campaigns work to cast doubt on climate science and to make fossil fuel products look sustainable when they really aren't, all while also shifting responsibility for climate solutions solely to individuals.

There are lots of tricks, traps, and communication strategies deployed to deceive when it comes to climate change. Unfortunately, a big part of the burden falls to us to be critical consumers of information. We also need to advocate for the values we want reflected by businesses we support. For example, the #AdsNotFit2Print campaign, put together by nongovernmental organizations, grassroots and climate advocacy groups, and journalists, was created to pressure the *New York Times* to stop promoting fossil fuels in its advertisements—including in sponsored content it creates for companies like Shell, Exxon, and Chevron.

As a consumer, signing petitions and "voting" with your dollars are climate actions that can add up and influence companies you rely on for food, news, clothing, banking, and other critical services.

▸ **Will you keep a critical eye out for climate information coming from the fossil fuel industry?**

NAVIGATING THROUGH THE ANTI-CLIMATE CAMPAIGN

ENERGY PRODUCTION AND TRANSPORTATION

WHAT ARE THE DIRTIEST AND DEADLIEST SOURCES OF ENERGY?

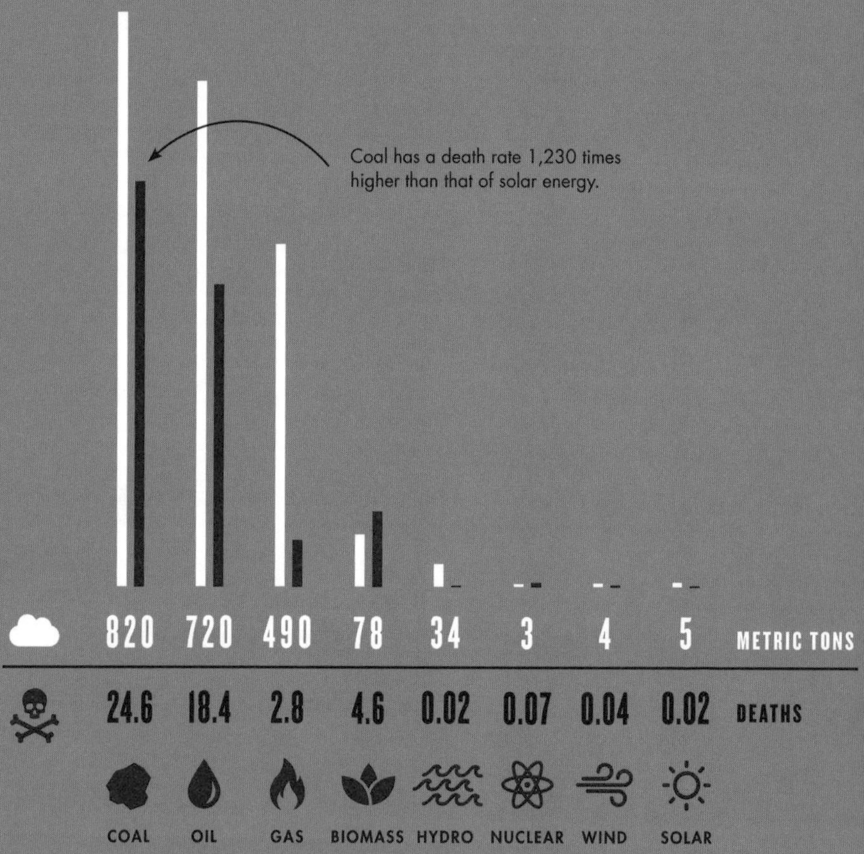

GREENHOUSE GAS EMISSIONS PER GIGAWATT HOUR OF ELECTRICITY OVER THE LIFE CYCLE OF THE POWER PLANT*

DEATH RATE BASED ON DEATHS FROM ACCIDENTS AND AIR POLLUTION PER TERAWATT-HOUR OF ELECTRICITY*

*1 gigawatt-hour is the annual electricity consumption of 150 people in the European Union.
1 terawatt-hour is the annual electricity consumption of 150,000 people in the European Union.

ACTION 6

KNOW WHAT POWERS YOU AND YOUR HOME

You know the adage, "knowledge is power," right? Part of how we each can decide where we can make the most difference when it comes to climate change is by researching and focusing on four things:

1. What climate impacts stem from our daily activities?
2. How can we contribute to climate action in our own communities?
3. Where do we have influence to motivate systems-based change?
4. What "wins" will motivate us to continue to act?

Let's look at home energy use as an illuminating example. Do you know what powers your appliances, lights, or heating system? Sure, your lightbulbs turn on with the flick of a switch and your blender whirs thanks to that outlet in your kitchen, but do you know what source of energy is creating the power to your house, condo, or apartment?

Finding out is relatively easy—for up-to-date stats, check out your energy provider's website or look elsewhere online. Many utilities also offer the choice of green energy options, like solar and wind power, to help support a greener, cleaner energy mix. For example, in Minnesota, the Minnesota Pollution Control Agency reported in 2021 that 75 percent of the energy consumed by Minnesotans was powered by coal, with only 1 percent from wind power. But many of the state's utility customers can buy green power to help support and accelerate the transition to clean energy for as little as $1.50 per month. In 2020, for an average Minnesota home, it cost $6 to $14 per month more to have 100 percent of its electricity come from renewable energy sources. This renewable energy is both greener and cleaner than coal.

By building your knowledge base and opting in to renewable energy, for example, you've understood your impact, you've contributed to climate action in your community and contributed to systems-based change, and hopefully this 'win' will fuel further climate actions. Now that's one action with a big impact!

▶ **Will you deepen your knowledge of what powers your home and community and opt into cleaner alternatives?**

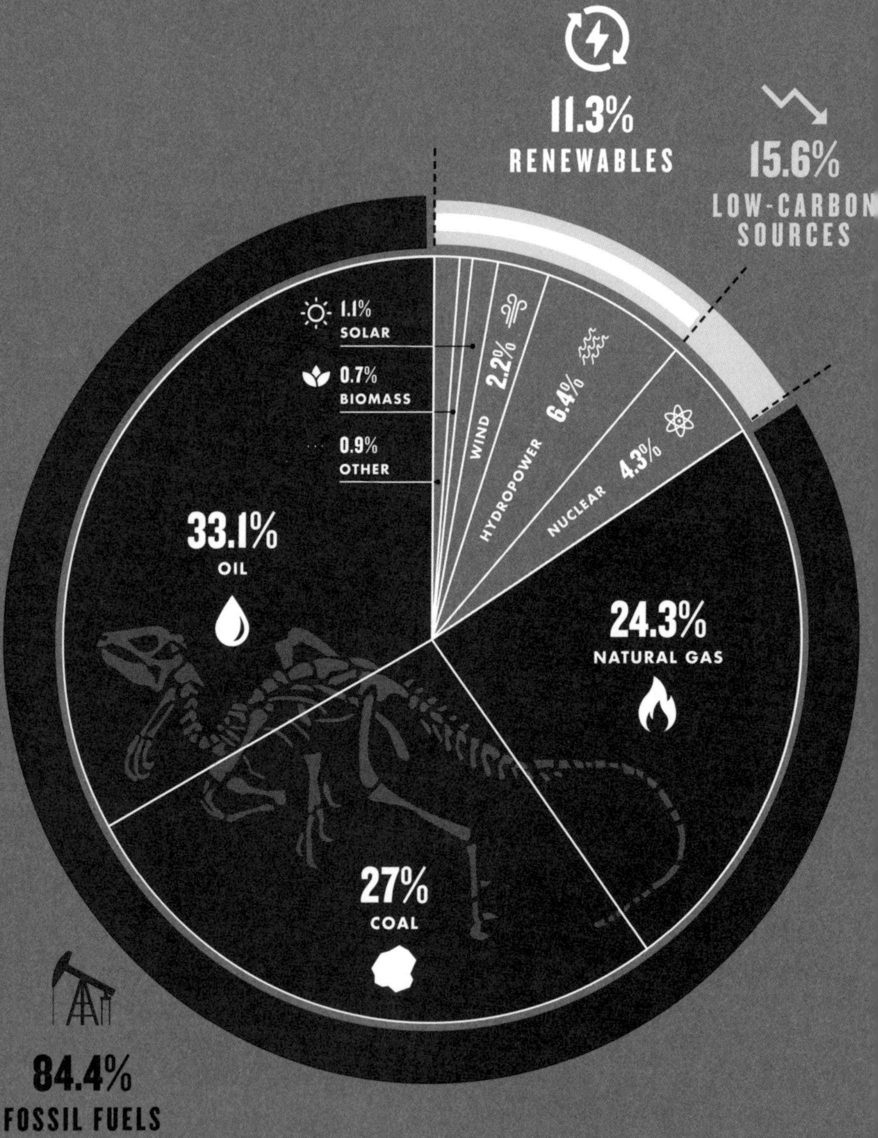

IN 2019, THE GLOBAL ENERGY MIX WAS STILL DOMINATED BY

OIL, COAL, AND GAS.

ACTION 7

CURB THE COST OF RENEWABLE ENERGY

Eighty percent of the world's energy still comes from fossil fuels, so how can we help accelerate the transition to cleaner energy? We know we need to expand the use of renewable energy like wind and solar power. But when it comes to expansion of these and other renewable technologies, the perception is that costs are astronomical compared to their fossil fuel counterparts. As it turns out, the costs of renewable energy continue to decline, while the profitability of high-polluting energy sources like coal may have already passed their peak. In fact, many new coal plant projects are being canceled, and since the adoption of the 2015 Paris Agreement, 44 countries (and counting!) have committed to not developing any new coal plants.

As we think about large-scale systems change, decreasing costs can be a significant milestone for turning the tide toward cleaner energy. Evidence indicates that as more and more renewable energy is being built and deployed around the world, costs are dramatically being driven down. In fact, in a 2020 analysis, the International Energy Agency (IEA) found solar photovoltaic systems in many areas are providing the cheapest electricity in history!

Looking for more good renewable news? Between 2009 and 2019, the cost of electricity from solar dropped a whopping 89 percent, while the cost of electricity from coal dropped only 2 percent! Other renewable energy sources are also getting cheaper—the price of onshore wind energy declined by 70 percent over the same ten-year period. This decline in costs puts both solar and wind as the cheapest sources of electricity on the market today. But wait—there's more! The cost of using batteries to store excess energy generated through these sources is also declining. Between 2012 and 2019, there was a 76 percent drop in the price of battery energy storage.

▶ **Will you look into the cost of clean energy in your community?**

ACTION 8

SUPPORT SOLAR AND OTHER RENEWABLES IN YOUR REGION

Do you know if your local utility uses alternatives to fossil fuels like solar, geothermal, wind, wave, or nuclear energy? Tools–like WeatherPower, by the nonprofit Climate Central–can help you explore the role your region is playing in the transition to clean energy, including calculating the portion of power being generated by solar and wind.

Shocked or pleased by what you see? Consider supporting renewable energy use in your state or in your own home. The cost of rooftop solar continues to fall, and more incentives are being made available for solar installations of all shapes and sizes. A sophisticated analysis by Project Drawdown demonstrates that rooftop solar is a critical climate solution, with anticipated global growth of gridded and stand-alone photovoltaic systems between 2020 and 2050 staged to avoid anywhere from nearly 28 to 69 gigatons of greenhouse gas emissions and saving anywhere from $7.89 trillion to $13.53 trillion in operational costs over the lifetimes of these systems.

In places where renewable energy is starting to be more widely used, send a note or social media post of support and gratitude to your neighbors who are installing systems on their homes and to your local utility letting them know you appreciate investments in a greener, cleaner future. We know we need a diverse portfolio of renewable energy options and the scaling of these solutions needs to happen at home and at large scales by utilities, governments, and corporations.

Want some inspiration? In 2018, 40 percent of installed solar photovoltaic capacity was located on rooftops, representing 25 percent of all renewable energy capacity globally. Even better, the potential to scale rooftop solar photovoltaic systems globally is significant. A 2021 study identified more than 77,200 square miles of rooftop area available for these systems globally!

▸ **Will you see if solar energy is an option for your home, school, business, or elsewhere in your community?**

GLOBALLY,

THERE ARE MORE THAN

77,200

SQUARE

MILES OF ROOFTOPS

AVAILABLE FOR

SOLAR

PANELS.

ACTION 9

WEIGH THE IMPACT OF DECARBONIZATION

There is no question that as we embark on a transition to greener, cleaner energy that we need to extract new, often virgin, resources to support this transition. Think metals and minerals needed for electric car batteries, solar panels, and wind turbines. This resource extraction is not free from its own negative environmental and social impacts.

The World Bank estimates that over 3 billion tons of minerals and metals are needed to expand wind, solar, geothermal, and energy storage capacity in order to meet the global goals to keep warming well below 2°C. This represents a 450 percent increase in mining and production of minerals like lithium, cobalt, and graphite by 2050 relative to 2018 levels. Unfortunately, mining and resource extraction has a legacy of leaving behind pollution and toxic waste, damaging landscapes, and negatively impacting the health of nearby communities and ecosystems.

Many of these activities occur in places with weak environmental regulations and labor safety practices. Therefore, it is critical to advocate for an ethical, just, and sustainable transition in all places where resource extraction occurs. This means supporting policies and regulations that proactively protect the environmental, social, physical, and cultural health and well-being of the communities sited near these extraction efforts. We also need to increase incentives and investments in practices and technologies that lead to improved efficiency of extraction and processing, and more recycling and reuse.

Concerns around a just transition are global in scale and require expanded investments, monitoring, and improved coordination across all scales of governance, both locally and globally. It can be worth the effort: even with the higher emissions associated with producing an electric vehicle (EV), the lower emissions from operating an EV over its lifetime compared to a conventional vehicle are proven. But we have to look beyond emissions to encompass the full environmental and social impact of this needed transition.

▸ **Will you help advocate to protect the environment and communities near extraction sites of these needed materials for decarbonization?**

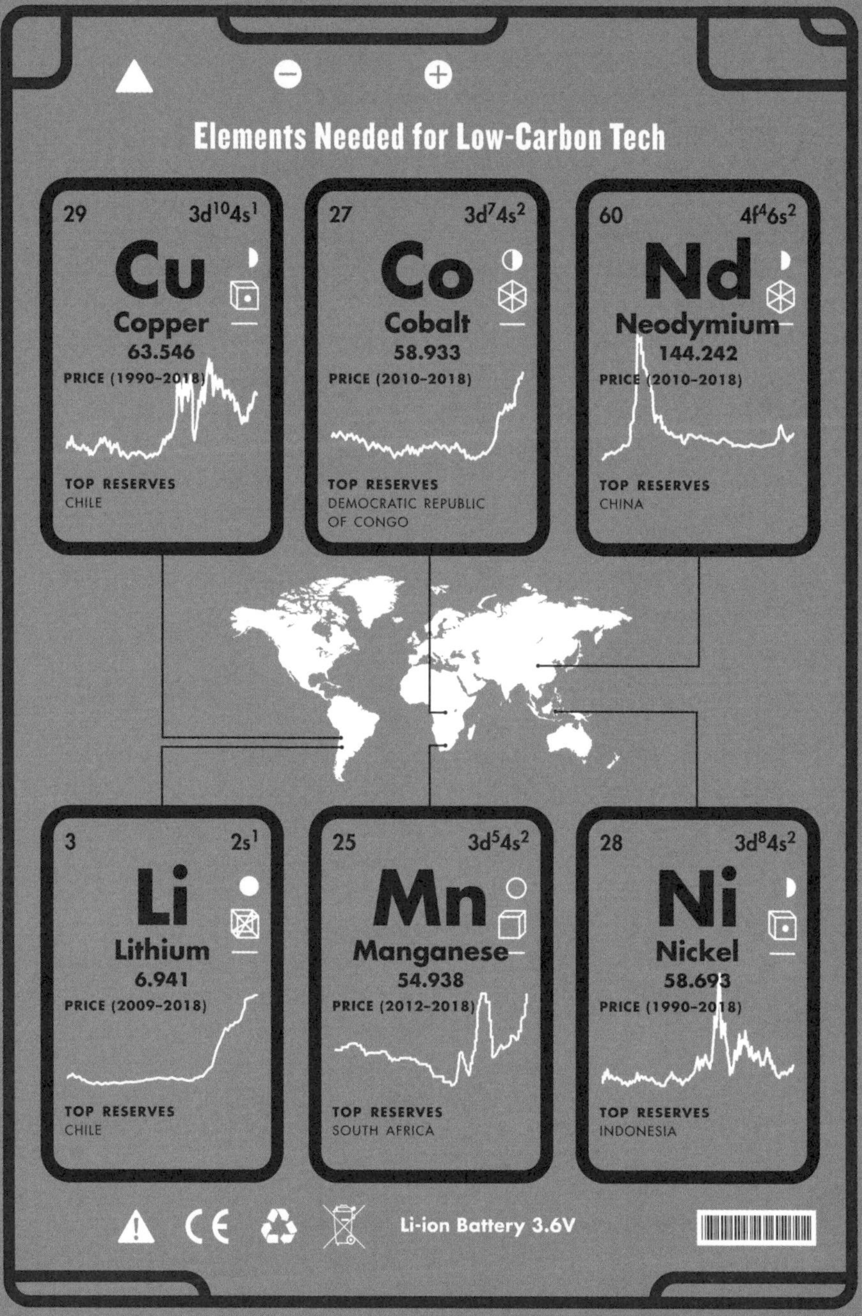
Elements Needed for Low-Carbon Tech
29
$3d^{10}4s^{1}$
Cu
Copper
63.546
PRICE (1990–2018)
TOP RESERVES
CHILE
27
$3d^{7}4s^{2}$
Co
Cobalt
58.933
PRICE (2010–2018)
TOP RESERVES
DEMOCRATIC REPUBLIC OF CONGO
60
$4f^{4}6s^{2}$
Nd
Neodymium
144.242
PRICE (2010–2018)
TOP RESERVES
CHINA
3
$2s^{1}$
Li
Lithium
6.941
PRICE (2009–2018)
TOP RESERVES
CHILE
25
$3d^{5}4s^{2}$
Mn
Manganese
54.938
PRICE (2012–2018)
TOP RESERVES
SOUTH AFRICA
28
$3d^{8}4s^{2}$
Ni
Nickel
58.693
PRICE (1990–2018)
TOP RESERVES
INDONESIA
Li-ion Battery 3.6V

ACTION 10

DRIVE AND COMMUTE MINDFULLY

In 2017, a Federal Highway Administration survey estimated that about 87 percent of Americans drove to work in private vehicles while only 6.9 percent used public transit as a "usual means of travel to work." This is a habit that has been sustained for decades. With transportation as a major source of emissions globally, a transition in what we drive, how far we drive, and the extent we use public transit or other nonautomotive forms of transportation (think bikes!) can all have a big impact.

While you may be thinking that it is too hard to replace car trips with other forms of transit, consider this: In the United States, 35 percent of vehicle trips are under 2 miles. Even if we replaced just one of those short trips each week with walking, we could save just over 4 gallons of gas each year. Walking not an option? Opting to use a bike for that same trip would yield the same benefits! No bike? No problem. Taking a bus would save 3.4 gallons of gasoline over the course of a year. All for swapping out just one 2-mile trip each week.

Seem trivial? Those numbers are for one person. Now just imagine if every American—all 332.5 million of us (and counting!)—opted for another form of transportation for just one 2-mile trip each week? The number becomes staggering.

Concerned about your commute to work and want to see how the numbers stack up? Check out the Get Around Greener tool on the Breaking Boundaries website to explore how your different commuting options can add up to real impact.

When it comes to climate change and transportation, there is no real conundrum. Public transit, walking, and biking are better for the climate and the environment. Unfortunately, many communities are not yet designed for alternatives to private vehicle travel, so we need to not only think about our own transportation patterns, but we also need to engage in efforts to increase access to public transit, electrify transportation systems like buses and ferries, and add more bike lanes to make a diversity of transportation options safe, accessible, and affordable.

▸ **Will you be more mindful of your commuting choices?**

DID YOU KNOW THAT SWITCHING FROM DRIVING AN AVERAGE SEDAN TO USING PUBLIC TRANSPORTATION CAN REDUCE EMISSIONS ANYWHERE FROM 26 TO 76 PERCENT?

—Wynes and Nicholas 2017, Environmental Research Letters

ACTION 11

CONSIDER CARPOOLING AND RIDESHARES

Reducing the number of individual car trips we make can have a real impact. If public transit, walking, or biking aren't regular options for you, consider ridesharing and carpooling. When using rideshare services like Uber or Lyft, opting to share with other passengers or carpooling with neighbors or coworkers can translate into significant emissions reductions. In fact, the climate solutions research organization Project Drawdown puts expansion of carpooling high on the list of actions we can engage in as individuals to reduce emissions. The organization's extensive research suggests that an expansion of carpooling globally could save from 4.17 to up to 7.7 gigatons of carbon dioxide emissions over the next 30 years (from 2020 to 2050).

Unpacking this further, in 2018 the global average occupancy of a vehicle was estimated to be 1.57 persons per trip. To achieve the 7.7 gigatons reduction in emissions, the average persons per trip would need to increase to 2.0 by the year 2050.

With technological advancements and increased access to smartphones, ridesharing is increasingly feasible. However, hopping in an Uber or Lyft by yourself and calling it a win for climate won't cut it. Without expansion in the number of passengers per vehicle, research suggests there are real drawbacks to shifting from private vehicles to rideshares. While rideshare services can lead to a reduction in vehicle pollution, like particulate matter and nitrous oxide, fuel usage and carbon emissions can actually increase from cars running continually. Non-climate-related factors, including noise, increased vehicle congestion, and the potential for more car accidents, also appear to worsen.

As with all climate solutions, context and coordination matter. Carpooling and ridesharing can translate into real benefits, particularly with fleets of zero-emission vehicles and when the number of passengers per car increases.

▸ **Who will you carpool and rideshare with?**

A three-person family always driving together instead of using their three family cars individually would save more than

3.2 METRIC TONS OF CARBON DIOXIDE,

and over

US $2,000,

in a year.

—Project Drawdown

GLASS MUST BE FULL BEFORE AND AFTER DELIVERY

IF THE UNITED STATES WANTED TO MOVE TO A FULLY

ELECTRIC FLEET

BY 2050

THEN SALES OF

GAS-POWERED VEHICLES

WOULD LIKELY HAVE TO

END ALTOGETHER

BY 2035

TAX INCLUDED

MAXIMUM ACCURACY AT ANY RATE OF DELIVERY AT ANY PRESSURE

ACTION 12

BUY AND DRIVE AN ELECTRIC CAR

In 2019, the transportation sector accounted for the largest fraction of total US greenhouse gas emissions, with 58 percent of those emissions produced by light-duty vehicles, including cars, vans, SUVs, and pickup trucks. Despite the fact that vehicles with conventional internal combustion engines are anywhere from two to four times less efficient than vehicles with electric motors, EVs account for only about 1 percent of the vehicles on the road today. Fortunately, increasing policy support for EVs, reduction in the cost of batteries, increased driving ranges, expansion of needed charging infrastructure, and more car manufacturers producing EVs are all helping to drive growth globally in both the passenger and commercial electric vehicle markets. In 2021, General Motors announced that by 2035 it will stop selling new gasoline-powered cars and light trucks. A 2020 study projected that by the year 2050, about half of the cars on the road around the world could be electric, saving around 1.5 gigatons of carbon dioxide emissions per year.

But are EVs actually greener if the source of the electricity that powers them is derived from fossil fuels? In short, yes. While there are some nuances, extensive "life cycle" analyses of the impacts of electric vehicles suggest that in 95 percent of the world, driving an electric car is better for the climate than conventional vehicles. While we need to carefully consider the cost of the raw materials required for EV batteries and expand ways to recycle batteries after their useful life in a car, the transition to cleaner cars needs to start today. Vehicle purchases today will play a big role in determining the composition of vehicles on the road for years to come, yet only 39 percent of US adults reported in 2021 that they were "very or somewhat likely" to consider buying an electric vehicle.

▸ **When the time comes for you to replace or upgrade your car, will you go electric? When you do, don't forget to look for incentives! Many EVs come with state and federal tax incentives.**

ACTION 13

DRIVE EFFICIENTLY

As we work toward vehicle electrification over the coming decades, we have options today to increase the efficiency of vehicles already on the road. We can all play our part. In 2019 the transportation sector in the United States accounted for 29 percent of the nation's total greenhouse gas emissions. With 58 percent of those emissions coming from vans, SUVs, and pickup trucks, how we use those vehicles matters. Maximizing efficiency, improving fuel economy, driving a small car, and reducing miles driven can mean big benefits for our climate and our wallets.

The US Department of Energy and the Environmental Protection Agency joined together to provide a range of options and suggestions for increasing fuel economy and trip efficiency. Have a look at some of their top suggestions:

- **REDUCE YOUR ROAD RAGE.** Aggressive driving can lower your gas mileage anywhere from 15 to 30 percent on the highway and up to 40 percent in stop-and-go traffic.
- **COMBINE SMALL ERRANDS INTO ONE BIG TRIP.** Short trips can consume double the fuel when compared to longer, multistop trips.
- **LIGHTEN THE LOAD.** Avoid hauling around extra weight in your trunk or on top of your car. Rooftop cargo carriers can reduce fuel economy up to 25 percent when traveling at highway speeds.
- **DIAL BACK THE CLIMATE CONTROL.** Blasting the heat or air-conditioning can reduce your fuel economy, requiring more fossil fuel use.
- **SLOW YOUR SPEED!** Fuel economy usually decreases at speeds above 50 miles per hour. Driving the speed limit, especially at top speeds, can make a big difference. For example, if you drive a 2017 Subaru Outback, you could save an estimated 31 cents per gallon by driving 65 instead of 70 miles per hour.

As noted in Action 12, we urgently need to shift to more electric vehicles and reduce overall car use. Not only can these actions reduce emissions but they can lead to improvements in air quality, cost savings, and—for those who swap sedentary transportation like driving for more active options—health benefits!

▸ **Will you reduce your speed, combine trips, and commit to driving less?**

@#$%*!

ENGAGING IN ROAD RAGE!
Lowers gas mileage by 15-30%

COMBINE SMALL ERRANDS INTO ONE TRIP
Reduces fuel costs by 50%

USING ROOF-TOP CARGO CARRIER
Reduces fuel economy by up to 25% when traveling at higher speeds

DRIVE EFFICIENTLY

SPEED LIMIT 65

DRIVE BELOW 65 MPH
Saves $0.31 per gallon

A/C

USING AIR-CONDITIONING
Reduces fuel economy by up to 25%

GO

ACTION 14

BE IDLE-FREE

You may have seen signs outside public buildings or schools in your neighborhood declaring them "idle-free zones." As it turns out, there are climate costs from letting our car's engine run when we aren't driving—say when you are waiting to collect kids at school, warming up your car on a cold winter morning, or waiting in the drive-through for coffee. Idling cars, trucks, and commercial vehicles can produce plenty of harmful emissions, including carbon dioxide and nitrous oxide and fine particulate matter that can reduce air quality. Did you know that idling for more than 10 seconds uses more fuel and creates more CO_2 than turning off your engine and restarting it?

Despite the fact that new cars and trucks don't actually benefit from idling, an estimated 6 billion gallons of fuel is wasted each year from idling heavy-duty and light-duty vehicles. According to the Department of Energy, about half of that wasted fuel is produced by personal vehicles, adding an estimated 30 million tons of CO_2 to the atmosphere each year. That's the same as the greenhouse gas emissions from about 5.9 million gasoline-powered passenger vehicles driven for a year. Consider this: just letting our cars run while we chat on the phone, wait for our kids, pick up a coffee, or [insert your favorite waiting-in-your-car activity] is both wasteful and a pretty simple and cost-saving climate action we can implement.

Many schools and businesses are taking note, and plenty of resources are available to advance idle-free campaigns at your own school or workplace. Many states and cities are taking notice too. In California, Hawaii, New Hampshire, and New York City, idling is actually illegal. Some car manufacturers are making it even easier with an "auto idle stop" feature that automatically shuts down and restarts a car's engine when you stop to reduce the amount of time your engine spends idling. Being idle-free is good for your wallet, your local air quality, and the climate.

▸ **Will you commit to being idle-free?**

Idling vehicles contribute to air pollution and emit air toxins known or suspected to cause cancer or other serious health effects. Monitoring at schools has shown elevated levels of benzene, formaldehyde, acetaldehyde, and other

AIR TOXINS

during the afternoon hour coinciding with parents

PICKING UP THEIR CHILDREN.

—US Environmental Protection Agency

TRAVEL AND WORK

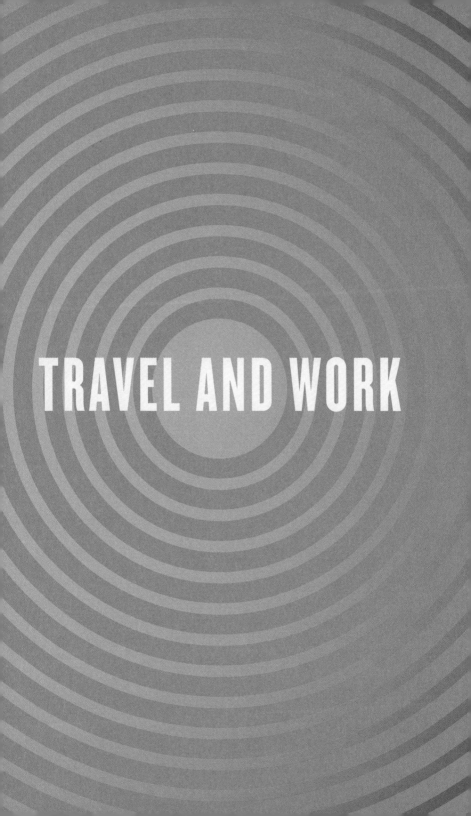

The Uneven Distribution of Regional Air Transport Demand

Measured in Revenue Passenger Kilometers (RPK)*

Year	Region	RPK Share (%)
2018	AFRICA	1.8%
2050	AFRICA	2.4%
2018	ASIA-PACIFIC	32.5%
2050	ASIA-PACIFIC	44.1%
2018	CIS**	2.5%
2050	CIS**	1.9%
2018	EUROPE	22.7%
2050	EUROPE	16.7%
2018	LATIN AMERICA	6%
2050	LATIN AMERICA	7.3%
2018	MIDDLE EAST	6.4%
2050	MIDDLE EAST	8.4%
2018	NORTH AMERICA	25.6%
2050	NORTH AMERICA	16.9%
2018	REST OF WORLD	2.5%
2050	REST OF WORLD	2.4%

5%
Average Annual Growth Rate of World Passenger Air Traffic

* RPK = total number of paying passengers X the distance traveled in kilometers

**The Commonwealth of Independent States (CIS) includes Armenia, Azerbaijan, Belarus, Kazakhstan, Kyrgyzstan, Moldova, Russia, Tajikistan, and Uzbekistan.

ACTION 15

FLY LESS, FLY ECONOMY

Commercial and freight aviation produces an estimated 2.5 percent of total global carbon dioxide emissions and 1.9 percent of total global greenhouse gas emissions. However, aviation's impact on the warming of the planet is higher, at around 3.5 percent, due to a complex set of factors including the production of water vapor, soot, and sulfate aerosols, and increased cloudiness that is created by the formation of contrails. While these numbers may seem small, between 2013 and 2018, aviation-based emissions increased by 32 percent, with an increase of around 5 percent in global aviation-based CO_2 emissions each year between 2010 and 2020. Aviation emissions, including all of the greenhouse gases emitted by airplanes, have doubled since the mid-1980s.

Despite growing global demand for flying, commercial aviation emissions are produced by a tiny fraction of the world's population, with only 11 percent traveling by air in 2018. Just 1 percent of the global population is responsible for 50 percent of the CO_2 emissions from commercial aviation. In 2018, the United States had a reported 590 million domestic air passengers, and North America accounted for over 25 percent of total global air transportation.

Even while there are new commitments from the aviation industry to expand the use of more sustainable fuels and create zero-emissions aircraft, those technologies are not yet proven at scale and are still a ways off from widespread adoption. So, for those of us who fly, flying less is an important climate act. In fact, flying can represent a disproportionate part of an individual's own carbon footprint. Mile for mile, flying is one of the most emissions-intensive activities many of us engage in. Flying in the front of the aircraft in premium classes results in an impact about three times worse—seats are larger, meaning fewer passengers can share the total emissions produced during a trip. If your work has traditionally required travel, options are increasingly available to limit it, such as virtual and hybrid meetings or conferences. For a more detailed look at the role of business and conference travel, check out Action 19. Considering the role of remote work? See Action 22.

▸ **How can you minimize your air travel, and will you sit in coach?**

ACTION 16

VACATION CLOSER TO HOME

Tourism creates and sustains jobs and drives economic growth, but it also comes with an environmental cost. Globally, tourism is an economic engine and supports jobs, with one in 10 new jobs related to the industry. Recent research indicates that up to 8 percent of global greenhouse gas emissions are related to the tourism industry, with an estimated 75 percent of related emissions tied to the transportation we use during travel.

With anticipated growth of the industry, forecasts indicate that tourism-related emissions will climb at least 25 percent by the year 2030, alongside an increase in both domestic and international air travel. Understanding the emissions associated with different forms of transportation can help us unpack our impact—and encourage critical thought about the best path to that next beach vacation, cruise, mountain getaway, or family reunion.

The math comparing emissions of train, plane, boat, and automobile travel can be complicated when considering the distance traveled, fuel sources, the fare class, and other details. At a high-level, however, the choices are relatively simple to boil down. Air travel produces significant emissions and for those who fly, is often an individual's most emission-intensive activity. By and large, travel by train, boat, or car is less carbon intensive. While it can sometimes be difficult to opt for an alternative mode of transit for travel, even cutting back on one plane trip a year can make a difference. Consider this: it takes 2 acres of healthy forest in the United States one year to capture and store the equivalent of the emissions created by each passenger on a round-trip transatlantic flight!

▸ **Will you consider a different mode of vacation travel or opt to vacation closer to home?**

MILE FOR MILE, FLYING IS THE MOST DAMAGING WAY TO TRAVEL FOR THE CLIMATE

EMISSIONS FROM DIFFERENT MODES OF TRANSPORT

PER PASSENGER PER KILOMETER TRAVELED

IN GRAMS OF CARBON DIOXIDE EQUIVALENT

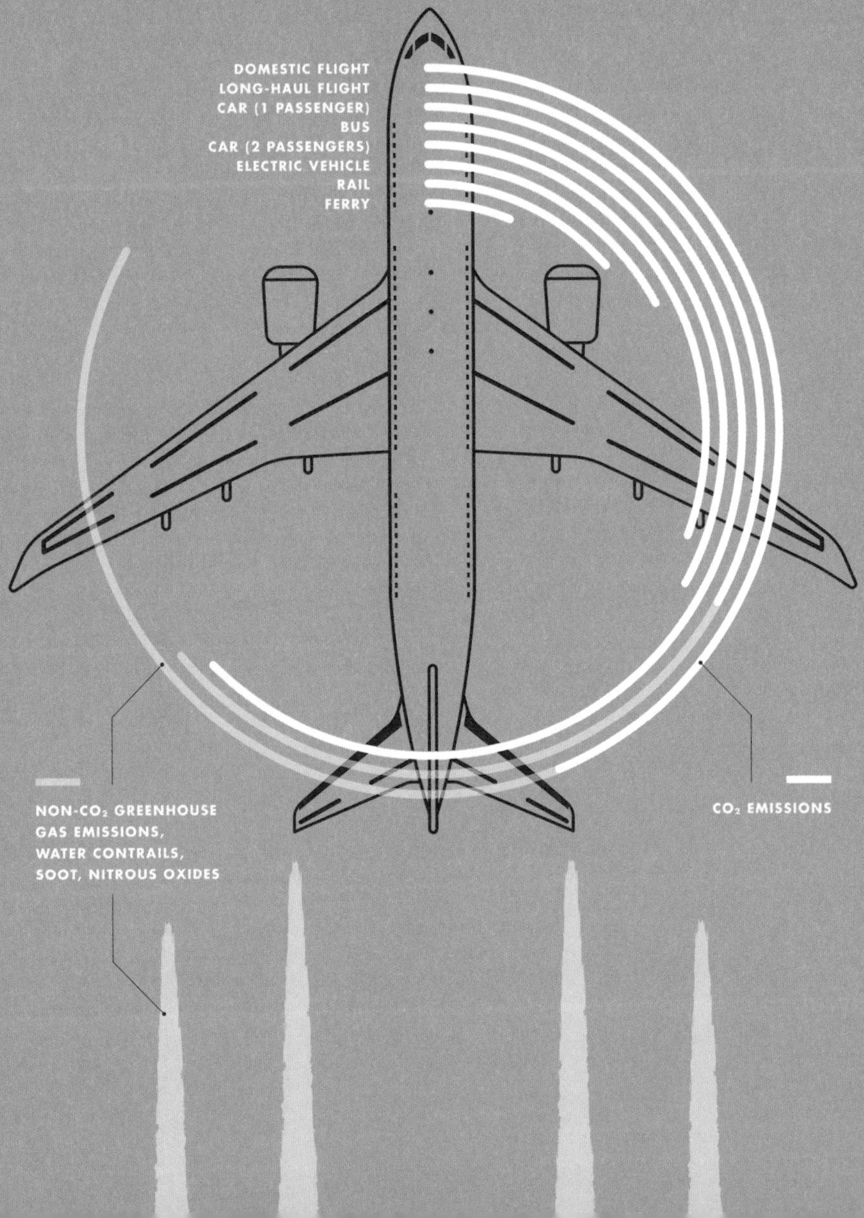

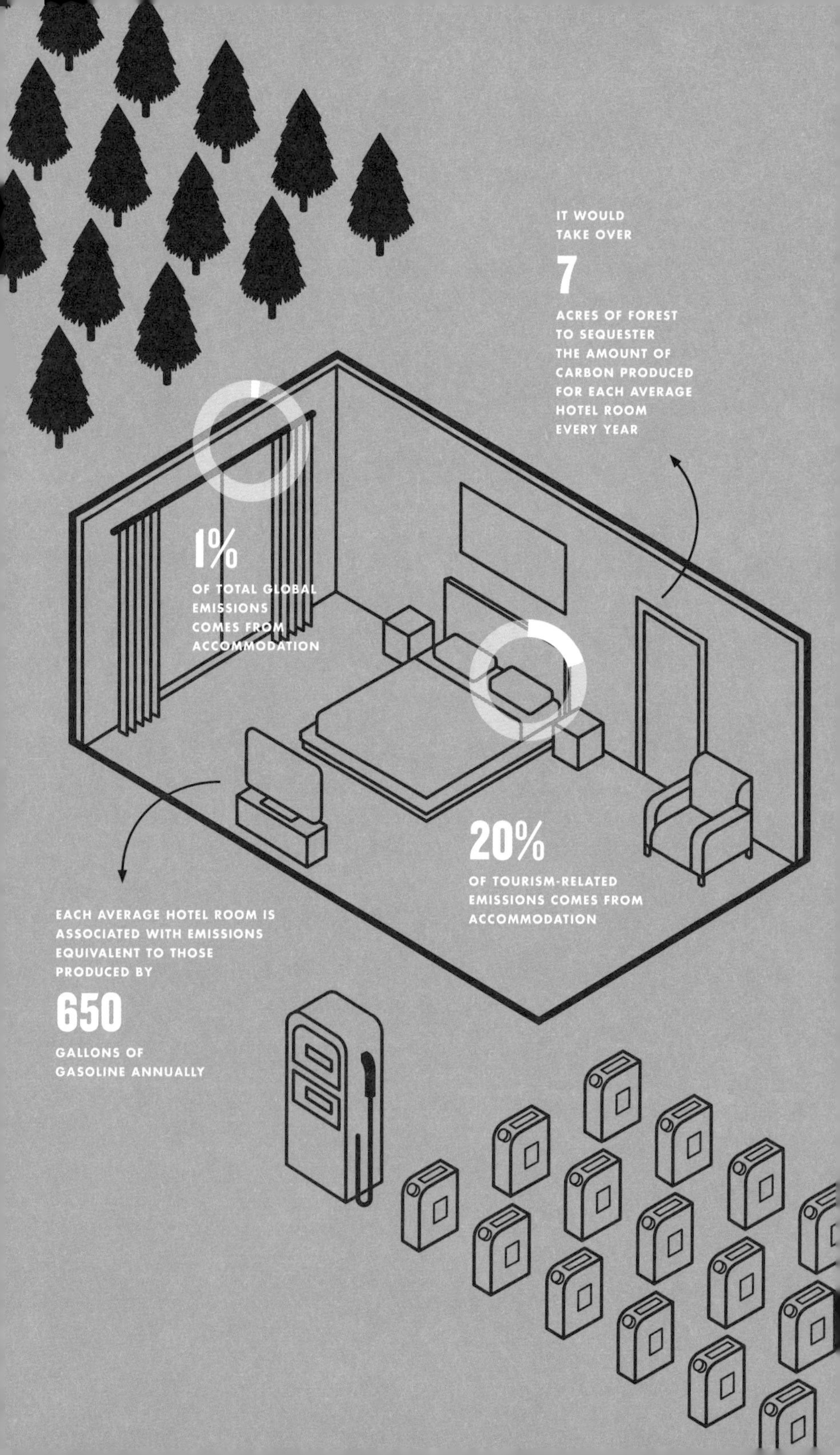
IT WOULD TAKE OVER
7
ACRES OF FOREST TO SEQUESTER THE AMOUNT OF CARBON PRODUCED FOR EACH AVERAGE HOTEL ROOM EVERY YEAR
1%
OF TOTAL GLOBAL EMISSIONS COMES FROM ACCOMMODATION
20%
OF TOURISM-RELATED EMISSIONS COMES FROM ACCOMMODATION
EACH AVERAGE HOTEL ROOM IS ASSOCIATED WITH EMISSIONS EQUIVALENT TO THOSE PRODUCED BY
650
GALLONS OF GASOLINE ANNUALLY

ACTION 17

SEEK OUT ECO-FRIENDLY ACCOMMODATIONS

While three-quarters of carbon dioxide emissions from tourism are related to transportation, there's no question that overnight accommodations for business and leisure travel come with an environmental cost. Accommodations, like hotels or motels, account for about 20 percent of tourism-related emissions and 1 percent of total global emissions annually. An average hotel room is associated with emissions equivalent to the amount produced by 650 gallons of gasoline annually. It would take over 7 acres of forest to sequester the amount of carbon produced for each average hotel room every year. Estimates suggest these emissions will continue to rise without the adoption of more sustainable practices and standardized emissions tracking and reporting. Research from the Sustainable Hospitality Alliance found that hotels need to achieve a whopping 90 percent per room reduction in emissions to be consistent with Paris Agreement targets to limit warming to below 2°C above pre-industrial levels.

What are the greener options for accommodations? A recent analysis suggests that the emissions per room per night are about the same whether you choose a home share, like an Airbnb, or an economy hotel room. Whether you choose home share or hotel, do your homework to make sure you choose accommodations that are transparent in their emissions reporting and invested in supporting initiatives that reduce water use and waste, or choose a host who prioritizes climate-friendly characteristics at their property, including use of renewable energy, composting, and water reduction options like low-flow appliances or rainwater harvesting.

While accommodations are an inevitable element of travel, you can still reduce the climate impact of your adventures by investing your resources in hosts or hotels that recognize their environmental footprint and are actively seeking to reduce their impact and are supporting their local communities.

▸ **Where are you going to stay during your next trip?**

ACTION 18

REDUCE TRASH WHEN YOU TRAVEL

It's way too easy to create more waste than usual when we travel, be it a road trip, a business trip, or a camping and hiking adventure. More to-go cups, individually packaged items, water bottles, plastic utensils, small toiletry containers, laundered towels and sheets, lack of recycling at hotels: the list is seemingly endless. One study suggests that tourists produce about twice the amount of trash than local residents each day. This usually includes single-use plastics. Hotel chain IHG reports that it uses 200 million miniature bottles of soaps, shampoos, and lotions every year. Marriott International reported an estimated 500 million tiny toiletries go to the landfill each year across its global properties! To address this waste, both chains have transitioned away from tiny toiletries to large refillable pump-top bottles. For Marriott, that translates to an estimated 30 percent reduction in plastic usage and about 1.7 million pounds of plastic waste avoided each year. Not all brands are taking the initiative, but in California they will soon have to; the state has enacted a law that will ban tiny toiletries in hotels by 2024.

Not only can our travel-related trash stress local waste management systems, but much of the waste we produce during travel can be avoided or minimized through a few simple steps: Bring your own toiletries in reusable containers or use solid soaps and shampoos. Bring your own water bottle and coffee mug. If you've opted for a homeshare, consider cooking instead of takeout. The main point is to be conscious of your consumption and any waste you are creating.

Consider making a trash travel log the next time you travel, to see where you might reduce waste. You can also assess your accommodations' environmental policies, including zero-waste toiletries and energy-efficiency upgrades, and opt to support those whose operations align with your (environmental) values.

▸ **How much trash do you create when you travel? What are opportunities for improvement?**

YOUR TRAVEL CHECKLIST

- Carry a refillable water bottle or mug
- Bring your own toiletries
- Seek out recycling and compost bins
- Limit use of single-use items
- Walk or take public transportation
- Open windows and turn off the AC
- Wear clothing more than once
- Hand-wash and air-dry clothing
- Research your lodging's policies
- Cook instead of dining out
- Use the same towels for your entire stay

ACTION 19

FIND ALTERNATIVES FOR WORK-RELATED TRAVEL

Work-related travel is big business, but it also has a big climate impact. Business spending on travel in 2019 exceeded $334 billion, with airfare costs accounting for about 20 percent of these dollars. While business travelers make up a small percentage of total air travelers (around 12 percent) they often provide a majority of the revenue for a given flight.

While there are, of course, many merits to in-person interactions and meetings, and positive impacts on the local economies and the service industry that supports the travel industry, there have been increasing calls across a range of professional societies and other groups to reduce travel due to the negative environmental impact of such travel. For example, a professional society of earth and space scientists holds a large annual meeting that gathers tens of thousands of people from dozens of countries. The estimated carbon footprint of air travel to this meeting is 15,000 metric tons, the same emissions produced by the electricity used by 2,725 average households for an entire year!

With conferences and trade shows constituting a large proportion of work-related travel, and accounting for more than $139.3 billion in travel-related spending in 2019, an expansion of virtual and hybrid conference options has the potential to both increase participation and reduce the environmental impacts of these gatherings. Of course, the COVID-19 pandemic altered the picture. The pandemic accelerated a transition toward less travel and a wider adoption of virtual meetings and more hybrid options for conferences. In 2020, companies reduced their travel budgets by more than 90 percent. In 2021, several large companies announced plans to reduce business travel over the longer term, with some of them anticipating travel-related emissions reductions of around 70 percent.

While most folks likely won't cut travel out completely, moderation can come with a range of benefits for people and the planet.

▸ **Will you opt to fly less for work? Will you encourage your company to make the switch to more virtual or hybrid meetings?**

Direct spending by resident and international travelers in the United States averaged

$3.1 BILLION

a day,

$128.5 MILLION

an hour,

$2.1 MILLION

a minute and

$35,700

a second.

—UStravel.org

ACTION 20

LEARN MORE ABOUT CLIMATE FINANCING

Funding and resources used to support the transition to renewable energy and to help communities prepare for and respond to the impact of climate change are often referred to broadly as climate financing. This financing comes from the public sector, corporations, banks, institutional investors, funds, and households. In 2009, developed nations that are part of the United Nations Framework Convention on Climate Change committed to "jointly mobilize $100 billion per year by 2020 in support of climate action in developing countries." Unfortunately, as of 2021, those commitments have not yet been met. In fact, the United Nations reported that between January 2020 and March 2021, more money globally was spent on fossil fuels than on investments in renewable energy and sustainable infrastructure.

Despite growing private and public commitments, climate financing from across a portfolio of public and private funds and intermediaries reached an average $632 billion by 2020. However, an estimated $4.35 trillion is needed each year over the next decade for global "energy systems, buildings, industry, transport, and other mitigation and adaptation solutions," according to the advisory organization Climate Policy Initiative. That represents a needed 590 percent increase from current climate financing. To put that another way, according to the experts, there is now an annual $3.5 trillion investment gap that needs to be bridged in order to deploy sufficient mitigation and adaptation solutions to meet our current global climate policy goals.

While governments and intergovernmental organizations, corporations, private funds, investors, and banks are needed to drive this large-scale climate financing, individuals can still play a role. Average 2019/2020 household and individual climate-related spending—which counts toward climate financing—totaled $55 billion. These climate-related dollars were primarily spent on electric vehicles, solar panels, and solar water heaters, and these investments are helping drive a transition to clean energy. Individuals are also increasingly seeking out sustainable investment opportunities to direct resources to companies and funds that provide climate financing for a range of climate-related projects. In fact, in 2019, 85 percent of individual investors were interested in sustainable investing, while 95 percent of millennials wanted to engage in sustainable investing. Despite this high level of interest, sustainable investing products and options have largely not kept pace with demand; in 2019, 65 percent of investors noted the lack of available investment products as a key impediment.

As with many of the actions in this book, when we are able to, it's essential to put our money where our climate values are. In this case, opt to support and finance renewable energy and decarbonization by purchasing an electric vehicle or adding solar panels to your home. This is a critical form of climate financing. Fortunate enough to be an investor or have an employer-sponsored retirement account? Invest in alignment with your values and talk to your advisor about available impact investing strategies that will help put your resources to work for climate solutions. We can also use our voices and votes to drive the change we want to see from our employers, financial advisors, banking institutions, or elected officials. With just one in 10 of us calling our elected representatives about climate change, you can make a difference by picking up the phone or sending an email to express your desire to see an expansion of climate financing locally, regionally, nationally, and globally.

▶ **How will you help to accelerate climate financing and investment?**

85% OF INDIVIDUAL INVESTORS ARE INTERESTED IN SUSTAINABLE INVESTMENT OPTIONS.

ACTION 21

DIVEST AND REINVEST

Since 1988, a mere 25 corporations and state-owned entities were responsible for more than half of global industrial emissions. According to the 2017 CDP Carbon Majors Report, in 2015 "a fifth of global industrial [greenhouse gas] emissions were backed by publicly listed investment." Further, the world's 60 largest banks have provided an estimated $3.8 trillion in needed financing to the fossil fuel industry since the Paris Agreement was adopted in 2015.

Since 2011 there has been a growing fossil fuel divestment campaign, driven primarily by the Go Fossil Free campaign–championed by 350.org–and student activists. The aim is to "cut off the social license and financing for fossil fuels," primarily by getting universities and other institutions to pull all of their investments (e.g., stocks, bonds, and other investment funds) tied to the fossil fuel industry. This divestment campaign is also trying to damage the reputation of fossil fuel companies in an attempt to keep fossil fuels in the ground and drive investment toward a clean-energy economy. The organization TH!RD Act is also working to hold big banks accountable through a "Banking on our Future" campaign, which is targeted at banks whose loans are supporting and helping to expand fossil fuel infrastructure including pipelines, oil terminals, and fracking wells.

The fossil fuel divestment campaign has gained momentum and is credited with changing the narrative around climate change. By the end of 2021, 1,500 educational, religious, and other institutions with assets over $39.8 trillion had committed to divest from fossil fuels. As funds have moved from these heavy emitters, there is a call to reinvest these financial resources in funds that are more aligned with each organization's values, including those that promote clean energy, social equity, and more.

A counterargument to divestment is to use financial engagement to drive and hold businesses accountable for helping accelerate the transition to renewable energy. For example, by staying engaged, banks—which hold considerable power as lenders—can use contractual and legal frameworks to hold companies accountable for progress toward decarbonization. If you want to dive into the merits and intricacies of divestment and reinvestment, consider a thoughtful internet search. If you want to track divestments across a growing number of organizations and institutions, check out DivestmentDatabase.org. If you yourself have investments, audit your portfolio and consider putting your resources in the growing number of climate-friendly and "impact investment" opportunities (see Action 20). Also consider where you bank and find out how your pension is invested. It pays to be fluent in the financials of climate change and engage (and divest, reinvest, or invest) where you can.

▸ **Will you review your investments and build a portfolio that aligns with your values? Will you encourage your employers and financial managers to put your money in more climate-friendly investments?**

Investors in fossil fuel companies carry influence over one-fifth of industrial greenhouse gas emissions worldwide.

—CDP Carbon Majors Report, 2017

ACTION 22

TELECOMMUTE WHEN POSSIBLE

The COVID-19 pandemic created seismic shifts in how many work, including a rapid expansion in the number of people working from home. Many signs point to remote work being part of a larger cultural shift going forward. In fact, a University of Chicago report suggests that 37 percent of jobs in the United States are able to be conducted fully remotely. Of those who can work remotely, only one in five reported working from home before the pandemic. That number rose to 71 percent in 2020. In addition to remote work, business travel also took a serious hit, and all signs point to sustained reductions in business travel relative to pre-pandemic levels (see Action 19).

The shift to remote work in 2020 was associated with improvements in air quality in 37 cities, and the trend toward more remote work is sustaining these air quality improvements.

In addition to improvements in air quality, reductions in business travel can be a big win for climate. With about 30 percent of commercial air travel associated with business meetings, Project Drawdown has estimated that replacing these trips with remote meetings would reduce 1 to 3.8 gigatons of carbon dioxide emissions over 30 years. It could also save $1.2 to $4.4 trillion (in 2014 dollars) and over 100 billion travel hours!

Don't travel for work? While its climate impact is full of nuances—depending on your geographic location, how you power your home, and your employer's sustainability practices—research shows that teleworking generally decreases total emissions. Despite the emissions benefits, there is a clear divide in who can actually work from home: higher-income workers and those with higher levels of educational attainment are more likely to be able to work from home. As some look to more remote work, we also need to look to more climate-friendly policies across our transportation systems, expanded climate commitments from corporations, and systems-scale changes that include consideration of the health and well-being of all workers, such as occupational hazards that leave some workers more exposed to climate extremes, like outdoor workers.

▸ **Can you work with your employer to allow for telepresence at least part of the time?**

37%

OF JOBS IN THE UNITED STATES ARE ABLE TO BE CONDUCTED FULLY REMOTELY.

OF THOSE WHO CAN WORK REMOTELY

20% REPORTED DOING SO BEFORE THE PANDEMIC

71% REPORTED DOING SO DURING THE PANDEMIC

IF PEOPLE CAN WORK TOGETHER WITHOUT BEING IN THE SAME PLACE, THEY CAN AVOID A HOST OF TRAVEL-RELATED EMISSIONS.

1-3.8

GIGATONS OF CARBON DIOXIDE EMISSIONS REDUCED OVER 30 YEARS

THE SHIFT TO REMOTE WORK IN THE SPRING OF 2020 WAS ASSOCIATED WITH IMPROVEMENTS IN AIR QUALITY IN 37 CITIES.

SPRING 2020 DECREASE IN NITROUS OXIDE EMISSIONS IN CALIFORNIA

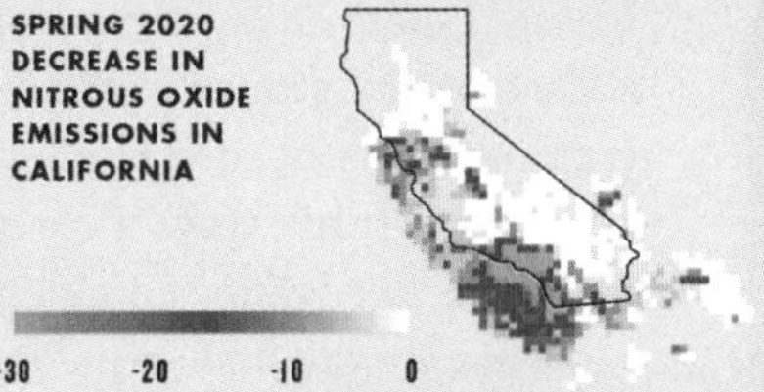

TELEPRESENCE

INTEGRATES VISUAL, AUDIO, AND NETWORK TECHNOLOGIES, SO PEOPLE CAN INTERACT ACROSS GEOGRAPHIES. IT CUTS DOWN ON TRAVEL—ESPECIALLY FLYING—AND ITS EMISSIONS.

30%

OF COMMERCIAL AIR TRAVEL IS BUSINESS RELATED.

27%

OF 2022 TRAVEL VOLUME IS PROJECTED TO BE REPLACED BY VIRTUAL MEETINGS.

GLOBAL CO_2 EMISSIONS DROPPED IN 2020

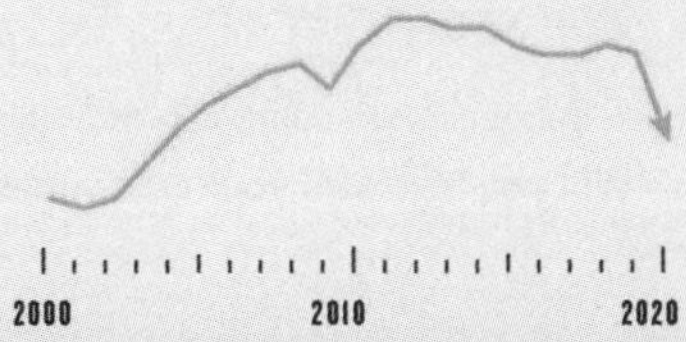

Unmute Stop Video Security Participants 10 Share Screen Chat Record Reactions End

ACTION 23

SEEK OUT CLIMATE SOLUTIONS IN THE WORKPLACE

Many of us spend a large part of our lives at work, and our jobs can be a big part of our identities. So shouldn't our companies and employers engage in climate work that we can all be proud of? Sound silly? Not only can actions by companies have a sizable impact on emissions reduction and in promoting and advancing climate adaptation, but research also suggests that companies engaged in climate change actions have better financial returns. Critically, climate-smart workplace cultures aren't just created or sustained by one person. Even if you have a designated sustainability director or leader, or a group dedicated to sustainability-related issues, efforts to "mainstream" climate change actions are needed across businesses of all shapes and sizes. Climate champions are needed across all business lines and across all groups of employees. Both business leaders and employees are needed to advance emissions reduction efforts and help identify ways a company can best use its influence and resources to advance climate policy, climate financing, and more!

What are the first steps? Do some research: What is your company or employer already doing to engage in climate solutions? Consider conducting an informal audit for climate opportunities in your workplace. The team at Project Drawdown created a very handy Climate Solutions at Work guide, which can help identify opportunities and inroads for solutions in the workplace.

To avoid being overwhelmed, this work is best done as a team sport. A perfect starting point can be a climate conversation with a colleague. From these sorts of conversations, you are likely to find a cohort of "climate colleagues" who can help champion change in your workplace.

▸ **What climate solutions and opportunities can you find in your workplace? Will you talk with your colleagues about climate solutions in your workplace?**

TEAMWORK MAKES THE DREAM WORK. IT CAN ALSO HELP TO ACCELERATE CLIMATE ACTION IN THE WORKPLACE.

61% OF THE ELECTRICITY USED TO POWER OUR BUILDINGS IS FROM FOSSIL FUELS WHILE ONLY 19.8% COMES FROM RENEWABLE SOURCES

ACTION 24

REDUCE THE CLIMATE FOOTPRINT OF THE BUILDINGS AROUND YOU

Buildings account for 39 percent of global energy-related emissions of carbon dioxide. About 11 percent of those emissions come from materials and construction, while 28 percent come from the regular operations of the building, including heating and cooling and lighting. Residential and commercial buildings account for 29 percent of US total greenhouse gas emissions when both the emissions from electricity generated off-site to power the buildings and direct emissions are accounted for. In 2015, the building sector was the fourth-highest-emitting sector in the United States.

In 2020, 61 percent of the electricity used to power US buildings came from fossil fuels. Clearly, there are major opportunities to make all buildings more climate friendly. Opportunities abound for greener buildings: better building design and standards, retrofitting and insulation, sustainable building materials, high-efficiency fixtures and appliances, smart thermostats, and increased use of renewable energy are just a few examples.

Globally, numerous policies and plans are underway to make buildings more climate friendly. New York City joined other smaller cities in phasing out all fossil fuel combustion in new construction. Commercial and residential buildings must also be fully electric by 2027. By 2040, this could prevent emissions equivalent to that produced by 450,000 cars in one year!

There is also a growing movement to make our buildings more resilient to the impacts of climate change. This includes ensuring adequate climate control and air filtration systems and placing a building's critical systems on higher floors to keep them out of harm's way from flood events.

If you own a home or business, consider improving the efficiency of the heating, ventilation, fixtures, and lighting. Install renewable energy systems like solar or geothermal, choose sustainable building materials, and replace appliances that use fossil fuels (see Action 54).

▸ **Will you work to improve energy efficiency in the buildings you work, live, and spend time in?**

ACTION 25

GO GREEN AND COOL WITH ROOFTOPS

Covering rooftops with plants and grasses spans back centuries, but green rooftops—and, increasingly, "cool roofs"—are gaining popularity for their clear climate and environmental benefits and cost savings.

Green roofs are created when a roof's surface is covered with living vegetation. Green roofs act to insulate a building, reducing the amount of heating and cooling required by the building, leading to lower energy use and a reduction in greenhouse gas emissions. Green roofs can also reduce stormwater runoff, provide habitat for birds and insects, clean the air, and, for many, offer aesthetic appeal.

Cool roofs are created when a reflective surface is added to a rooftop. They similarly reduce the amount of energy required to cool a building. The lighter-color surface of a cool roof reflects heat rather than absorbing it, like its darker counterparts, acting to reduce the internal temperature of the building by up to 30 percent and requiring less energy to keep the building cool and comfortable. Both green and cool rooftops in urban environments can also help reduce the urban heat island effect (see Action 72).

Want a cool cool-roof example? Take a look at New York City, which has had a dedicated cool-roof program called NYC CoolRoofs since 2009. This program created 10 million square feet of cool rooftops between 2009 and 2021, with an estimated reduction of one ton of carbon dioxide saved for every 2,500 square feet of cool roof in the city! The program even offers some cool rooftops free of charge—to nonprofits, community or recreational centers, hospitals, museums and cultural centers, affordable/low-income housing units, and more! Any building owner in New York City can also get a cool-roof coating at a reduced cost. Once purchased, building owners can get technical support, materials like rollers and paintbrushes, and costs of labor covered by the CoolRoof program. What's not cool about that?

Not in New York City? Other places are also embracing cool roofs, including Ford Field, a domed stadium in Detroit, and federal government buildings like the Department of Energy headquarters.

Contrary to what you might think, it isn't just commercial or flat rooftops that can go green or cool. The reflective materials used for commercial and larger-scale residential applications can also be applied to create cool roofs on private homes. Cities like Portland, Oregon, are actively encouraging residents to install green roofs through their eco-roof program, which includes a do-it-yourself guide and a planning guide.

Worried about cost? The EPA puts cool-roof materials at about 10 to 20 cents more per square foot than conventional rooftops, but research suggests cool roofs offer up to 50 cents of yearly net savings per square foot. You do the math!

Curious about a cool or green roof? A good starting point is the Department of Energy's Cool Roofs web page (Energy.gov/EnergySaver/Cool-Roofs), which outlines a number of options for existing roofing and new construction. If you aren't a homeowner, consider pitching the benefits of a cool roof to your building operator or owner. Or follow New York City's example and advocate for a coordinated cool-roof program in your community.

▸ **Will you consider installing or advocating for a green or cool rooftop?**

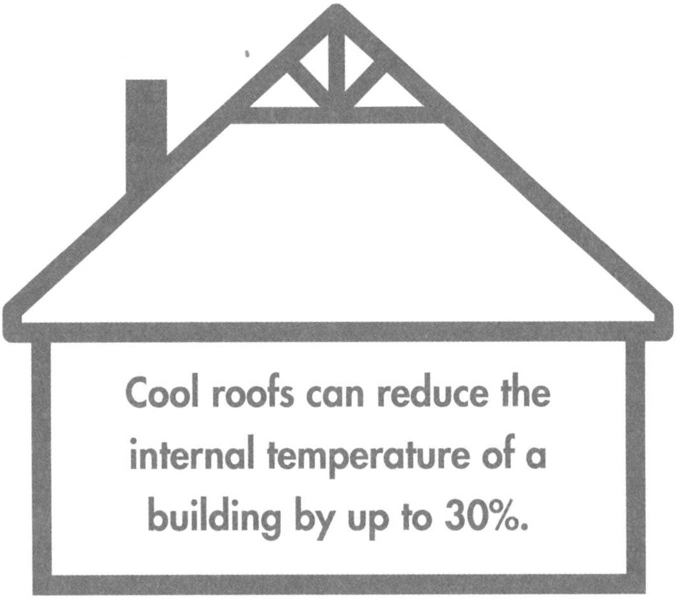

Cool roofs can reduce the internal temperature of a building by up to 30%.

ACTION 26

CONSIDER CLIMATE AS PART OF YOUR CAREER

Despite arguments that a transition to renewable energy will kill jobs, evidence suggests that, in fact, a range of climate jobs are expected to emerge in the coming decades. A 2022 study evaluating the impact of a transition from fossil fuels to renewable energy sources suggests that the uptake of renewable energy will increase jobs in the energy sector from about 57 million in 2020 to nearly 134 million in 2050. While some jobs will undoubtedly be lost, other sectors of the economy will experience job growth. Subsidies, expansion of training and education efforts to upskill or reskill workers, and increasing domestic manufacturing can help to accelerate and facilitate a just transition, in which jobs and economic gains are realized.

On the downside, there are risks to certain sectors, like tourism and hospitality. Workers are also expected to become less productive as the planet warms. Between 2000 and 2018, extreme heat resulted in a loss of nearly 1.1 billion potential labor hours, with large impacts on the agriculture sector. This loss of productivity, which includes the prospect of premature death for some workers, is expected to cost the United States billions of dollars in the coming years. However, with bold climate action, overall economic and human health gains are expected to be significant; the World Resources Institute suggests a shift to renewable energy and preventing worsening climate change could add up to 65 million new "low carbon jobs" by 2030 and up to $26 trillion in cumulative economic gains. In addition, 700,000 premature deaths could be avoided through associated improvements in air quality as we transition away from fossil fuels.

Even if your job isn't at risk, climate actions can show up in all of our careers—whether you are a small business owner, a stay-at-home parent, an engineer, or an artist. We can advocate for a transition to a low-carbon economy that seeks to ensure jobs and job transitions are supported across the multiple sectors that will either contract or expand as we confront climate change.

▶ **Will you consider how climate fits into your career and support a just transition?**

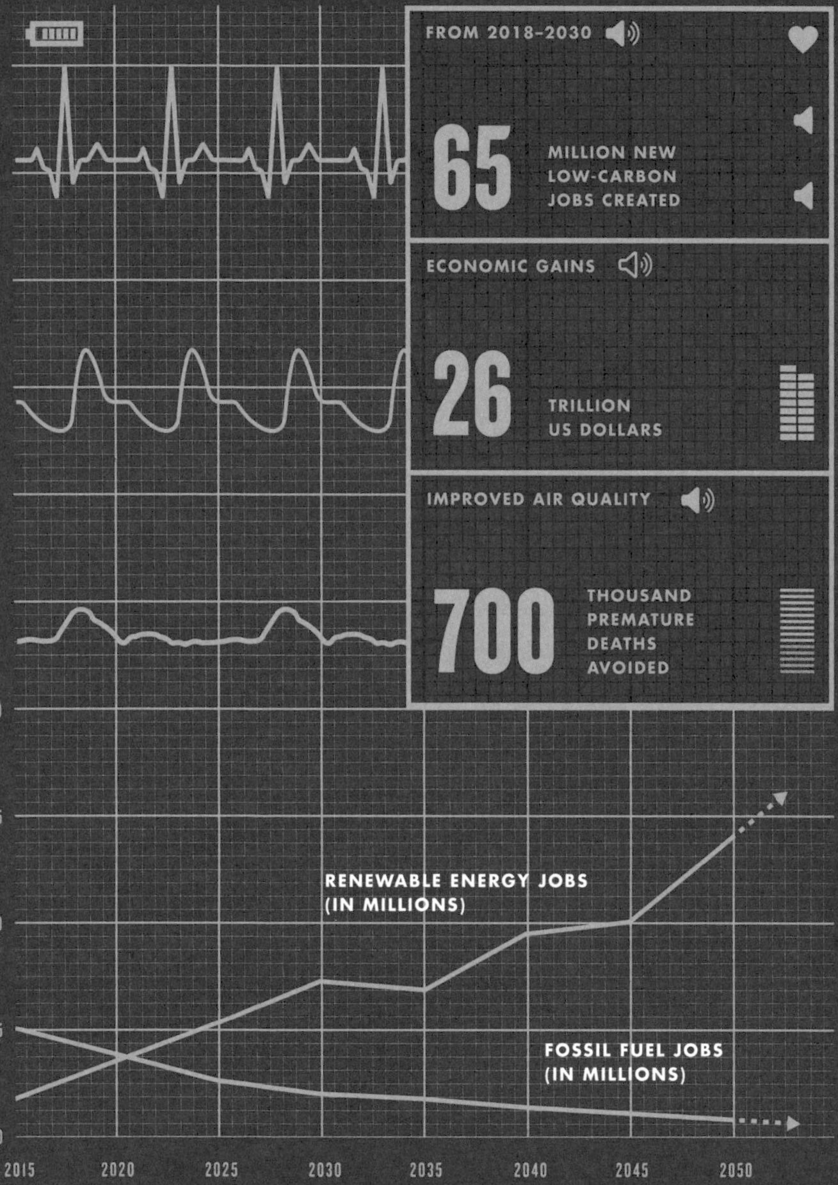

TRANSITIONING TO RENEWABLE ENERGY LEADS TO ECONOMIC AND HUMAN HEALTH GAINS

ACTION 27

USE CAUTION WITH CORPORATE CLIMATE COMMITMENTS

Increasingly, companies are making statements that claim they are working to reduce and defray emissions through carbon offsets in order to reach net zero. Net zero emissions is when the amount of greenhouse gases being created or emitted is matched by the amount being removed from the atmosphere. However, the quality of these climate commitments varies and regulation and accountability mechanisms are lacking. A healthy dose of skepticism and a keen eye for detail are needed when looking at a company's (or a government's) climate commitments.

That said, some companies have adopted rigorous climate commitments, and the number looking to is growing, with more than 1,000 companies setting science-based targets aligned with approaches needed to meet the 1.5°C (2.7°F) target set forth in the Paris Agreement.

While some of these commitments may seem impressive at first glance, many companies' commitments account for only the emissions produced by the company directly—say through powering its buildings or business travel—but not for emissions generated from its supply chain, including getting goods to customers. Estimates suggest that these supply-chain-related emissions can account for 80 percent or more of the emissions produced by a company.

If that's not confusing enough, many companies that purchase carbon offsets, which pay someone else to "offset" the company carbon emissions through undertaking a more environmentally friendly activity. This is often done through activities like planting a tree or supporting agricultural practices that help to store carbon in soils. However, these offsets alone are insufficient to meet our global climate goals. Many of these voluntary offsets are still not regulated, so it can be hard to track the quality or integrity of them.

One quick way to understand a company's commitments is to look for keywords like climate-neutral and offsets, and to look for a reliance on carbon capture and storage technologies, which are not yet proven or deployed at scale. If you see an over-reliance on such methods, you might want to take these claims with a grain of salt.

To create mechanisms and regulations for transparency and accountability, which would help consumers assess commitments (and subsequent follow-through!), government intervention is needed. Until then, we need to vote with our dollars and do our research to better understand if a corporate commitment is just idle talk or if the company is "walking the walk" when it comes to their climate commitments. Experts suggest that customer and employee pressure, including speaking out and engaging on social media, can work to increase corporate climate commitments and accountability.

▸ **Which companies will you hold to their climate commitments?**

OFFSETS can realistically do only so much for reaching climate targets. That is why the focus must turn toward REDUCING rather than offsetting global emissions.

—TheConversation.com

FOOD AND FARMING

CONSUME 25% LESS

BEEF
HIGH CARBON FOOTPRINT RESULTING FROM METHANE EMISSIONS AND LAND CONVERSION.

CO_2 EMISSIONS FROM PLANT-BASED PRODUCTS ARE 10–50 TIMES LOWER THAN ANIMAL-BASED PRODUCTS.

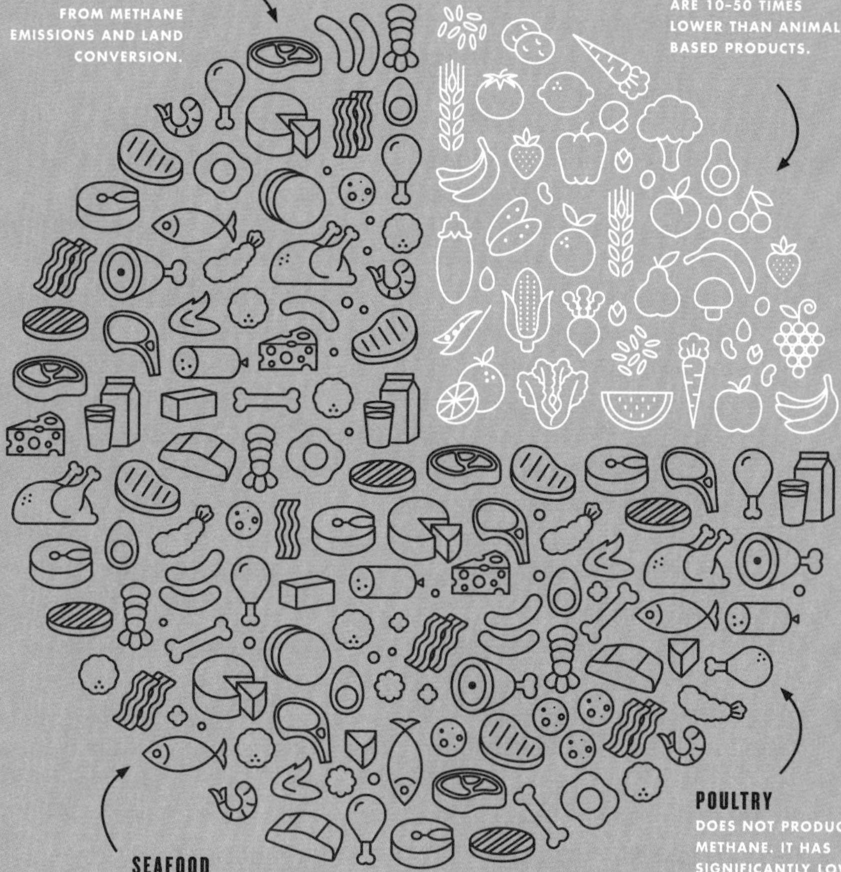

POULTRY
DOES NOT PRODUCE METHANE. IT HAS SIGNIFICANTLY LOWER EMISSIONS THAN BEEF OR LAMB.

SEAFOOD
COMPARATIVELY, FISH AND SHELLFISH HAVE A MUCH LOWER CARBON FOOTPRINT THAN LAND-BASED PROTEIN.

MEAT & DAIRY

ACTION 28

EAT YOUR BROCCOLI, AND PASS ON THE MEAT

Shifting our diet to one heavier in plants, rather than meat and dairy products, offers a meaningful opportunity for both mitigating and adapting to climate change. According to the Intergovernmental Panel on Climate Change, dietary changes, including adoption of a primarily plant-based diet, have a potential to mitigate 0.7–8.0 gigatons of carbon dioxide equivalent each year! To put that in perspective, you'd have to have 2,174,536 wind turbines running for a year or take 2,141 coal power plants off the grid for a year to avoid the same amount of emissions! Fortunately, to achieve these reductions, you don't have to forfeit all meat and animal products. Start by eliminating meat from one meal a week or cutting back meat and dairy consumption by about 25 percent. By serving smaller portions and replacing meat products with vegetables, grains, fruits, and protein-rich legumes like beans and lentils, you can reduce your climate impact and improve your health. It's a win-win.

Not only does shifting our diet have consequences for our health and carbon footprints, but shifting what we produce and grow on our farmlands can have an added environmental benefit—by enriching our soils, reducing erosion and runoff, and minimizing the need for fertilizer (which also has a carbon footprint!). Shifting what we grow and how we grow it can make our lands more able to handle the stressors of climate change, like the increasing extreme precipitation projected for much of the Midwest, one of the most productive agricultural regions in the United States. In fact, it is estimated that dietary changes alone could free up nearly 500 million acres of land for other uses by 2050, like planting trees and reforestation, which could further reduce the impacts of a changing climate.

▸ **What will be on your dinner table tonight?**

ACTION 29

CUT DOWN ON YOUR FOOD WASTE

In addition to striving to eat less meat and dairy (see Action 28), reducing food waste can reduce emissions of greenhouse gases. It is estimated that 40 percent of food in the United States is thrown away! In 2010, about 133 billion pounds of food available at retailers in the United States went uneaten, the equivalent of 141 trillion calories. Wasted food like this accounts for as much as 10 percent of global human-produced greenhouse gas emissions. The waste occurs against a backdrop of one in eight people in the United States struggling to put food on the table and two billion people globally lacking regular access to sufficient nutritious food in 2019. The United Nations estimates that without changes in policies and behaviors, the number of people affected by hunger will exceed 840 million by the year 2030—9.8 percent of the expected global population.

It's clear that addressing food waste comes with substantial climate and societal benefits. And it's important to note that, at present, consumers—not grocery stores or restaurants—are the biggest culprits when it comes to food waste in the United States. The upside of this is that we can clearly be part of the solution!

To reduce your food waste, you can invest more time in planning your meals, brush up on your food storage skills, and use scraps and leftovers in new recipes. (Make vegetable or chicken stock, for example.) Then be sure to compost any remaining food scraps to reduce your carbon footprint and put some of your food waste back into the soil. Flip the page to Action 30 to learn more!

▸ **How can you reduce your food waste?**

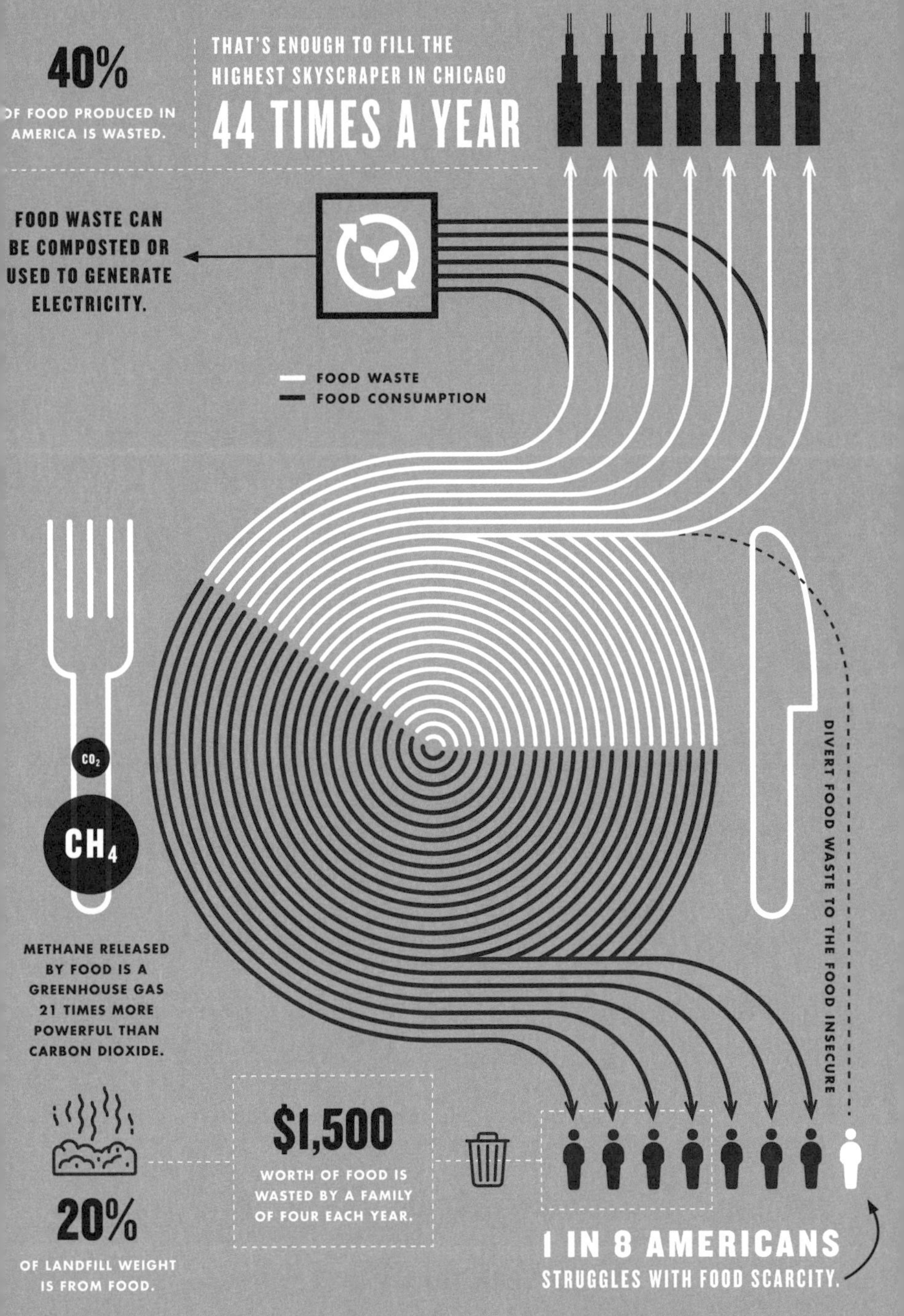
40%
OF FOOD PRODUCED IN AMERICA IS WASTED.
THAT'S ENOUGH TO FILL THE HIGHEST SKYSCRAPER IN CHICAGO
44 TIMES A YEAR
FOOD WASTE CAN BE COMPOSTED OR USED TO GENERATE ELECTRICITY.
FOOD WASTE
FOOD CONSUMPTION
CO_2
CH_4
METHANE RELEASED BY FOOD IS A GREENHOUSE GAS 21 TIMES MORE POWERFUL THAN CARBON DIOXIDE.
DIVERT FOOD WASTE TO THE FOOD INSECURE
20%
OF LANDFILL WEIGHT IS FROM FOOD.
$1,500
WORTH OF FOOD IS WASTED BY A FAMILY OF FOUR EACH YEAR.
1 IN 8 AMERICANS
STRUGGLES WITH FOOD SCARCITY.

WHAT TO COMPOST
BROWNS
GREENS
6
C
CARBON
12.011
7
N
NITROGEN
14.007
4.1%
OF FOOD WASTE IN THE UNITED STATES WAS COMPOSTED IN 2018.
COMPOST
25%
25%
50%
WATER
YARD WASTE + FOOD SCRAPS + MOISTURE + TIME
WHAT NOT TO COMPOST
LANDFILL/ INCINERATION
CREATE COMPOST
INDUSTRIAL USES
FEED ANIMALS
FEED THE HUNGRY
SOURCE REDUCTION
LEAST PREFERRED
FOOD RECOVERY HIERARCHY
MOST PREFERRED

ACTION 30

COMPOST

Did you know that a majority of the food that goes uneaten in the United States each year ends up in landfills, where it decomposes into the potent greenhouse gas methane? In fact, this rotting food accounts for a whopping 16 percent of US methane emissions and up to 10 percent of total global human-induced greenhouse gas emissions. Methane is a powerful heat-trapping gas, trapping about 100 times more heat per molecule than carbon dioxide. Curbing our methane emissions—produced predominantly through agricultural activities—offers a powerful way to make progress toward reducing overall planetary warming.

In addition to reducing food waste, composting food and yard waste offers a way to make an impact at home. Compost is created when yard waste (e.g., grass clippings, leaves, etc.), food scraps (e.g., eggshells, coffee grounds, onion skins, etc.), and moisture are mixed together and allowed to decompose. Over time, under the right conditions, this waste can produce a rich brown-black humus that can be added to your garden. It can take anywhere between two months to two years for compost to develop, but it can be worth the wait: not only does this process reduce methane emissions generated from food waste in landfills, but compost can also enrich your soil, enable it to retain more moisture, reduce pests and diseases, and eliminate the need for chemical fertilizers, among a long list of benefits. Along with the many options for backyard composting, a growing number of municipal programs directly collect compost as part of standard curbside waste services. Despite all of this, only 4.1 percent of our food waste in the United States was composted in 2018. Composting can be a low lift but high climate impact action!

▸ **Will you compost your food scraps and yard waste?**

ACTION 31

SHOP FOR YOUR MEALS MINDFULLY

Feeling guilty about grocery delivery or ordering those perfectly portioned meal kits? You might be surprised at how they stack up in terms of their climate impact. Recent research that assessed the full life cycle of meal delivery kits suggests that the climate calculus is actually in your favor, despite the higher use of plastic and disposable materials relative to goods purchased at a grocery store. On average, the emissions of delivered meal kits can be 33 percent lower than a typical "grocery meal." Meal kits tend to reduce food waste, which in turn leads to a reduction in emissions, even with higher packaging-related impacts, because waste tends to be a substantial chunk of the overall carbon footprint of our food.

Not into meal kits? Grocery deliveries can reduce the number of individual trips to the store, resulting in a reduction in transportation-related emissions. In fact, well-designed systems that cluster grocery deliveries by geographic location rather than timed windows can reduce emissions by 80 to 90 percent, compared to personal trips by car to the grocery store! Dense urban areas can benefit most from these emissions reductions, because companies can optimize delivery routes and reach more customers more efficiently.

As we look for solutions, we need to promote systems changes that encourage less packaging and prioritize the reduction of negative environmental impacts. We also need to voice our support for local and regional planning that would optimize grocery delivery routes in the future.

As individuals, we can combine as many grocery trips as possible to reduce transportation-related emissions, reduce food waste, reduce our consumption of heavily packaged foods, and choose meal kits made from locally sourced ingredients. Of course, the food in your orders matters! As discussed in Action 28, reducing consumption of meat and other carbon-intensive foods in your diet can also shift the carbon consumption math lower for your household.

▸ **How are you going to shop for next week's meals?**

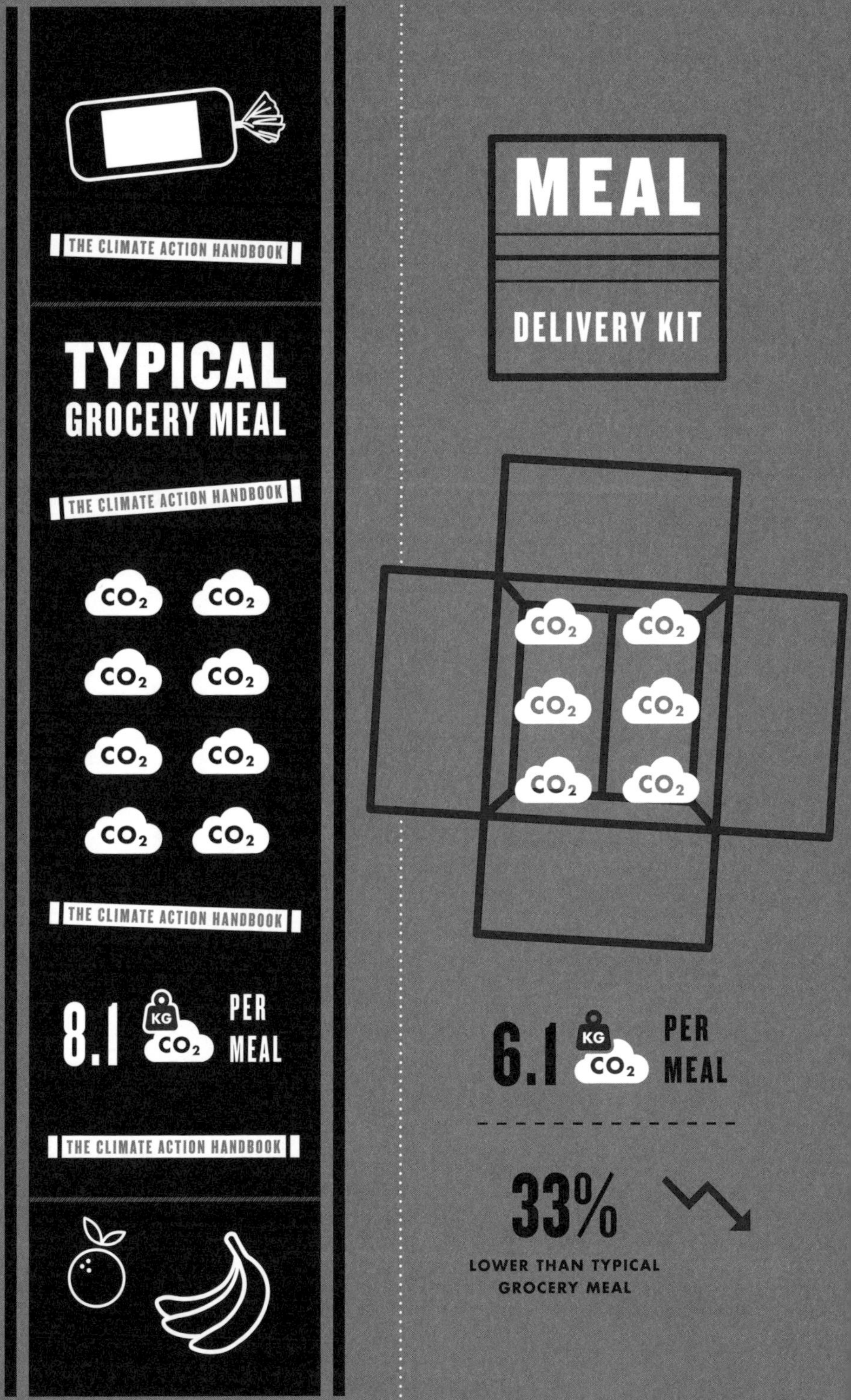
THE CLIMATE ACTION HANDBOOK
TYPICAL
GROCERY MEAL
THE CLIMATE ACTION HANDBOOK
CO2 CO2
CO2 CO2
CO2 CO2
CO2 CO2
THE CLIMATE ACTION HANDBOOK
8.1 KG CO2 PER MEAL
THE CLIMATE ACTION HANDBOOK
MEAL
DELIVERY KIT
CO2 CO2
CO2 CO2
CO2 CO2
6.1 KG CO2 PER MEAL
33%
LOWER THAN TYPICAL
GROCERY MEAL

WHICH TAKEOUT CUISINE PACKAGING PRODUCES THE HIGHEST EMISSIONS?

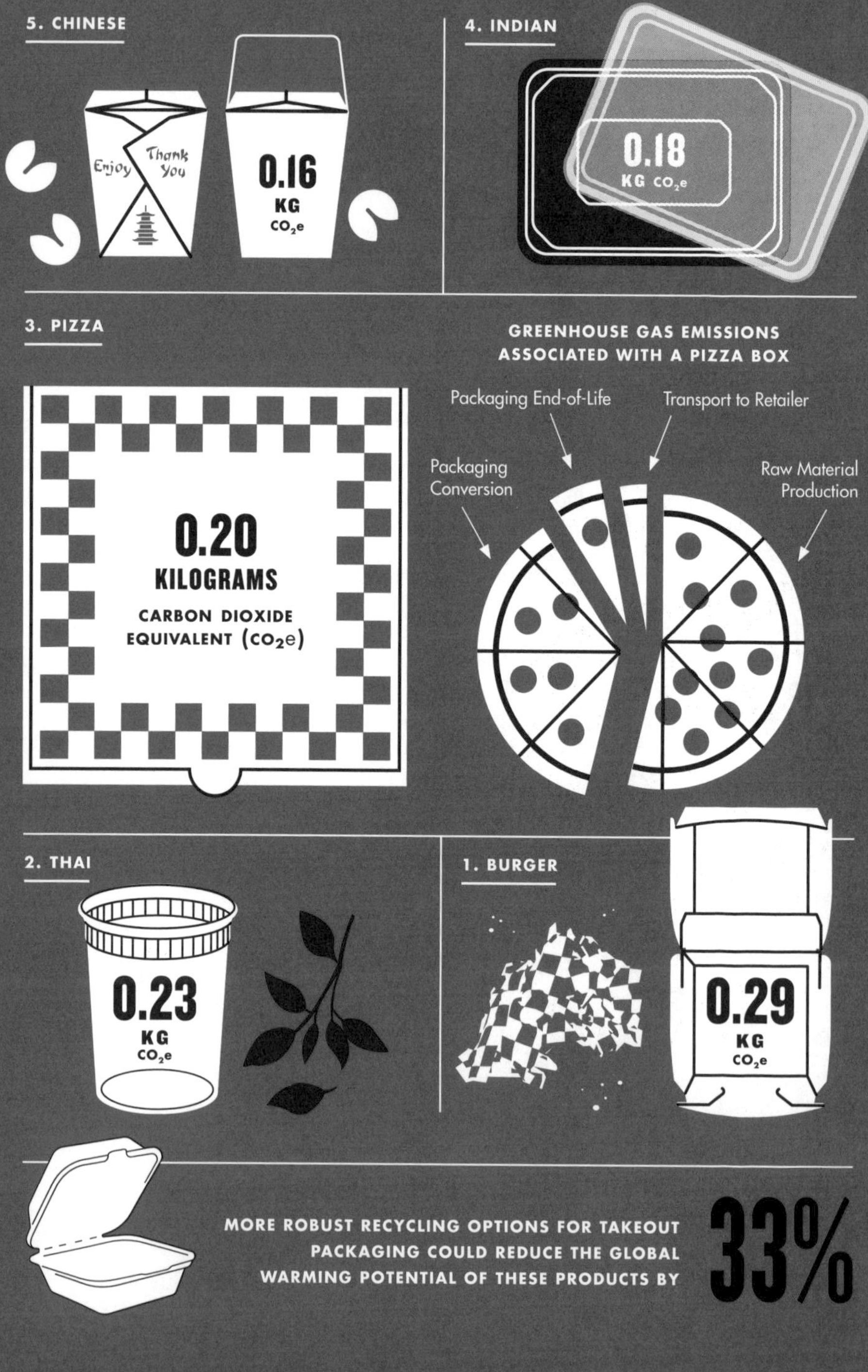

ACTION 32

RESEARCH MORE THAN THE TAKEOUT MENU

The COVID-19 pandemic caused food delivery app use to surge in the United States. In 2020, an estimated 45.6 million people used food delivery apps. Between April and September of that year, these apps brought in approximately $5.5 billion in combined revenue, which was more than double the revenue for the same period in 2019.

But does takeout pack an emissions punch? The answer depends on where you live and what you order, but on the whole, takeout and delivery generally come with a large carbon footprint, due primarily to the packaging, much of which can't be recycled.

An Australian study looked at the emissions associated with the packaging of five popular takeout foods—including pizza, burgers, and Indian, Thai, and Chinese cuisine—and found that burgers come with the most carbon-intensive packaging (not to mention meal choice!). Each year, the United States alone uses over 7.5 billion extruded polystyrene containers (think of those handy foam clamshell boxes), which represents 297 million metric tons of carbon dioxide emissions per year.

While there are different merits and climate impacts of different packaging options, a big part of the problem comes with the single use of the products and the lack of effective and comprehensive recycling programs. One study found that the implementation of more robust recycling options for takeout packaging across the European Union could reduce the global warming potential of these products by 33 percent.

While we can reduce our consumption, when we do opt for takeout we can do a few things to reduce our impact:

- Look for restaurants that use less, and more environmentally friendly, packaging. Believe it or not, there's an app for that! Jybe helps consumers find restaurants that use sustainable packaging.
- Be responsible disposing of your waste. Know your local recycling options, and if they don't yet accommodate the packaging from your favorite restaurant, advocate for change at the restaurant and with your waste management program.
- Finally, when you have the option to compost, please do.

▶ **Next time you order takeout, what will you consider besides the entrees?**

ACTION 33

THOUGHTFULLY OPT FOR MEAT ALTERNATIVES

Per capita meat consumption in the United States is three times the global average. However, plant-based food products that are designed to taste and feel like meat are increasingly popular, partly driven by an interest in health and the environment. In the United States alone, the plant-based food industry is expected to grow nearly tenfold by 2030.

A lesser-known alternative to meat is cultivated meat. Cultivated meat creates meat without an animal in a pasture. Instead, animal cells get replicated—or "cultivated"—in a lab. Start-ups in this space are increasing in number, and there are calls for greater investment and financing to accelerate growth of this industry.

But can we calculate the climate costs of plant-based and cultivated meat in comparison to their "traditional" meat counterparts? The team at the Breakthrough Institute did some of the heavy lifting on synthesizing the research and showed that, unsurprisingly, it depends! What determines the overall impact of cultivated meat is not only related to reductions in land use and nutrient pollution associated with not having animals in the field, but also what cultivated meat replaces in someone's diet. If all Americans replaced 40 percent of their beef consumption with cultivated meat, it could result in the equivalent of as much as 94 million fewer metric tons of carbon dioxide emissions. In contrast, choosing cultivated meat over chicken or pork might lead to increased emissions.

Scratching your head yet? The Breakthrough Institute puts it this way: "A diet including chicken and pork, but no dairy or beef, has lower greenhouse gas emissions than a vegetarian diet that includes milk and cheese, and almost gets within spitting distance of a vegan diet." The main messages are these:

- On the whole, plant-based foods are less resource intensive than animal protein. However, the footprint of plant-based meat or cultivated meat can be either large or small.
- Eating less beef, even if it is replaced with chicken or pork, translates to fewer emissions.
- As in life, everything in moderation.

▸ **How will you cultivate a more climate-friendly diet?**

A 40% SHIFT OF US MEAT CONSUMPTION TO CULTIVATED MEAT MAY NOT ALWAYS REDUCE EMISSIONS

Change in Emissions
(million metric tons CO_2 equivalent)

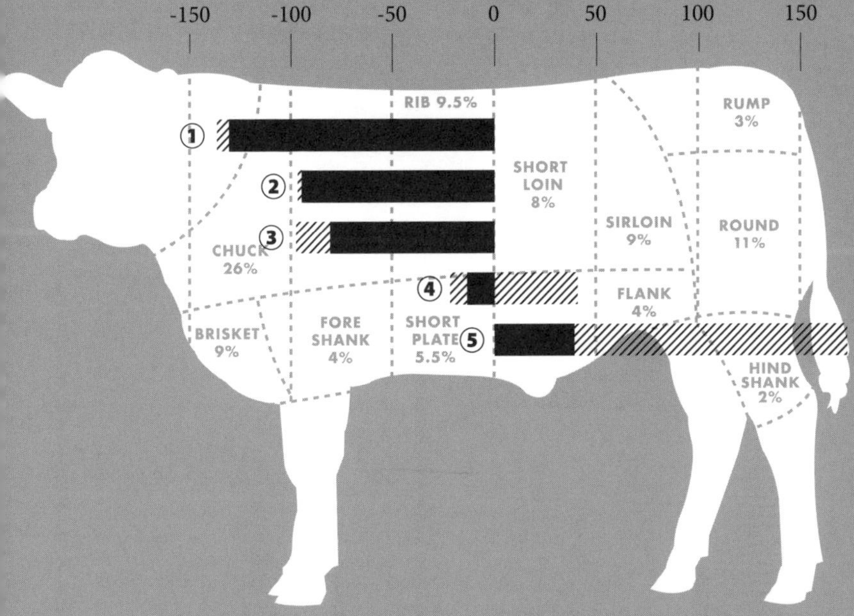

① Beef, Pork, and Poultry to Tofu
② Beef Burger to a Plant-Based Burger
③ Beef to Cultivated Meat
④ Pork to Cultivated Meat
⑤ Poultry to Cultivated Meat
Margin of Error

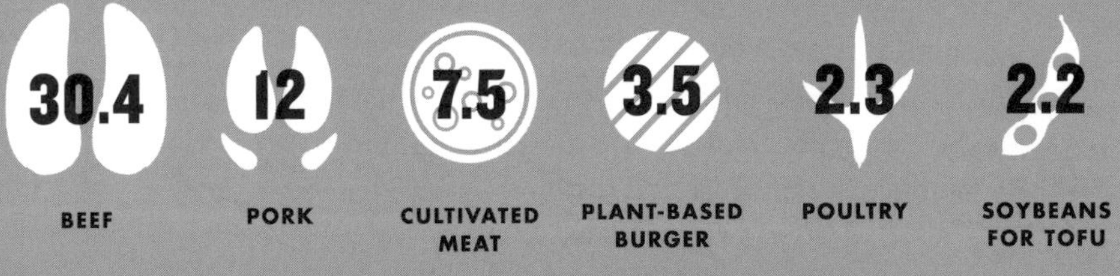

Carbon Footprint of Meat and Meat Alternatives
(kg CO_2 equivalents/kg product)

EMISSIONS FROM PRODUCING FOUR CUPS OF:

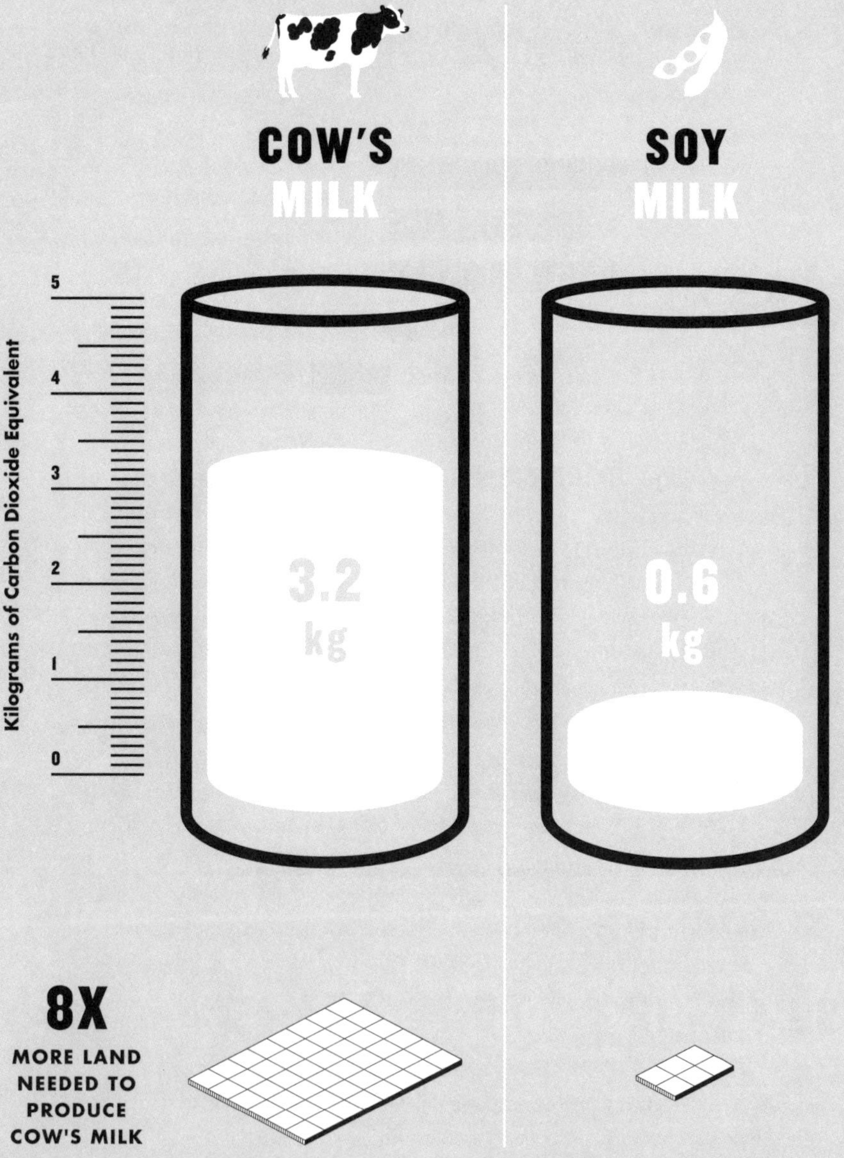

ACTION 34

SWITCH TO NONDAIRY ALTERNATIVES

Nondairy milks come in various forms, from soy and almond milk to oat milk and even quinoa milk! Worldwide sales of nondairy milk alternatives more than doubled between 2009 and 2015, reaching $21 billion in sales. Nondairy milks accounted for 13 percent of US retail milk sales in 2018. Though often made to look like cow's milk, the nutritional value of these milks can vary, and even with fortification some argue that they don't yet match the nutrition offered by cow's milk. For example, cow's milk is rich in protein, with around 8 grams of protein per 8 ounces, whereas soy milk contains around 1 gram of protein per cup.

Looking beyond protein and nutritional profiles, the climate impacts of milk and nondairy alternatives can be complex, with emissions depending on farming and land use practices, packaging, and distance between consumer and producer. For example, almond milk requires a substantial amount of water, fertilizer, and pesticides. Soy farming has also been associated with deforestation in places like Brazil, bringing with it questions related to the full environmental footprint of soy milk beverages. When it comes to nondairy alternatives, some experts suggest that oat milk might be the way to go, due to oat's relative abundance in production already and its ability to be grown in areas that could limit further deforestation.

On the whole, all of the food we consume comes with impacts. Globally, our food supply chain accounts for 26 percent of human-caused greenhouse gas emissions, with the growth and production of food accounting for well over half of food's emissions footprint. While there are impacts and trade-offs, experts suggest a few consumer practices can help limit the impact: avoid companies and producers with high negative impacts, opt for less packaging, and avoid wasting food.

Given that animal agriculture currently accounts for 51 percent of agricultural greenhouse gas emissions in the United States, and 16 percent of those emissions are associated with dairy, even a small reduction in total consumption of dairy products can reduce the climate impact of our food and beverage choices.

▸ **How can you reduce dairy consumption in your household?**

ACTION 35

LOOK INTO EATING LOCALLY

There are many motivations for eating locally—for environmental, social, economic, or health reasons, or because of animal welfare, labor practices, increased trust in local producers, and a sense of ethical or moral obligation. Many also perceive local food as being safer, tastier, and fresher and are willing to pay more for locally produced foods. However, the nuances of whether local food is "better" for the environment really depends. What is considered local; on-farm practices, like nutrient management and fertilizer and water use; food processing; and, ultimately, how food from any distance is transported to the consumer all influence the environmental footprint of a given food product. Typically, local food is considered to be food produced within a 100-mile radius of the consumer; however, studies suggest that food transport in the United States accounts for "less than 15 percent of the energy use and greenhouse gas emissions of food products."

Take, for example, the tomato—a 2013 study comparing processed tomatoes (diced tomatoes and tomato paste, to be exact!) consumed in Michigan, which were produced in either California or Michigan, found that "California-produced conventional and organic tomato paste and canned diced tomatoes are almost equivalent in energy use and [greenhouse gas emissions] to regionally produced and consumed products, but use of developed water resources is significantly higher for California-grown products." This same study showed, however, that the mode of transport of the food really matters—a shift from rail to truck transport for the tomato products completely changed the calculus of faraway food versus locally sourced food. A shift to truck transport resulted in a fourfold increase in energy use associated with the California-grown tomatoes.

One study summed it up nicely: "Food systems can never be reduced to a simple binary of *local is better* and *conventional is worse*, or its inverse *local is worse* and *conventional is better*, because of the complexities of the production and distribution systems and their relationship to [greenhouse gas] emissions."

So there are many ways to slice a tomato when it comes to the food sources and the values that motivate your decisions. Agricultural practices matter, as well as the energy used to power processing plants, as well as what environmental impacts are being considered. In addition to emissions, food production can bring a range of environmental impacts, including biodiversity and habitat loss, water use, nutrient-rich runoff that impacts water quality, and greenhouse gas emissions.

For consumers, knowledge really is power, and increased transparency about the different impacts and distances traveled—perhaps achieved by labeling—might help consumers make more informed decisions. A little research, and even some conversations with your local producers and other food providers at a place like a local farmers market, can go a long way in helping you weigh the different pros and cons of your food source choices.

▸ **What's in your pantry? Do you know where it came from?**

Food systems can never be reduced to a simple binary of *local is better* and *conventional is worse*, or its inverse, *local is worse* and *conventional is better*, because of the complexities of the production and distribution systems and their relationship to [greenhouse gas] emissions.

—Christensen et al., 2017

ACTION 36

ENJOY YOUR CHOCOLATE RESPONSIBLY

With more than 268 million people in the United States consuming chocolate or candy in 2020 and the average person in Switzerland consuming around 20 pounds of chocolate each year, it might be helpful to understand chocolate/climate change connections.

Cacao trees, which produce the seeds and pods that are a key ingredient in the chocolate many of us love, are kind of like Goldilocks from the fairy tale (a bit fussy—about their climate setting). Cacao trees prefer an environment with high humidity, consistent temperatures, lots of rain, and little wind. Their current "happy habitat" is only about 1,300 miles on either side of the equator. While cacao trees are grown in South America and Hawaii, 60 percent of cacao grown today is sourced from smaller farms in Côte d'Ivoire, Ghana, and Indonesia.

Climate changes stress cacao trees through increases in pest and disease risk, changes in the timing and characteristics of the wet and dry seasons, and, of particular concern, drier conditions. This drying can reduce the amount of water available to the trees, reducing yields. As such, research suggests that the optimal locations for cacao trees will change, and that more trees will be better able to survive higher up in the mountains, leading to concerns about conservation and changing land use patterns.

Increased farmer education, improved future climate projections, advanced preparation and land conservation measures, and the development of more climate-tolerant varieties of cacao are all paths toward a sustainable chocolate industry. In Peru, wild strains of cacao might prove to be more resilient to climate change, ushering a shift in the type of cacao grown and offering hope.

Consuming chocolate with climate change in mind can include supporting smaller producers who adopt sustainable practices and trying alternative varieties and sources of chocolate. A shocking amount of chocolate is also tossed every year so, as with other food, avoid buying more than you can eat. If you are curious about more climate and food connections, check out *Climate-Smart Food*, a recent book by Dave Raey. This well-researched book is available online for free!

▸ **What will you consider when buying chocolate?**

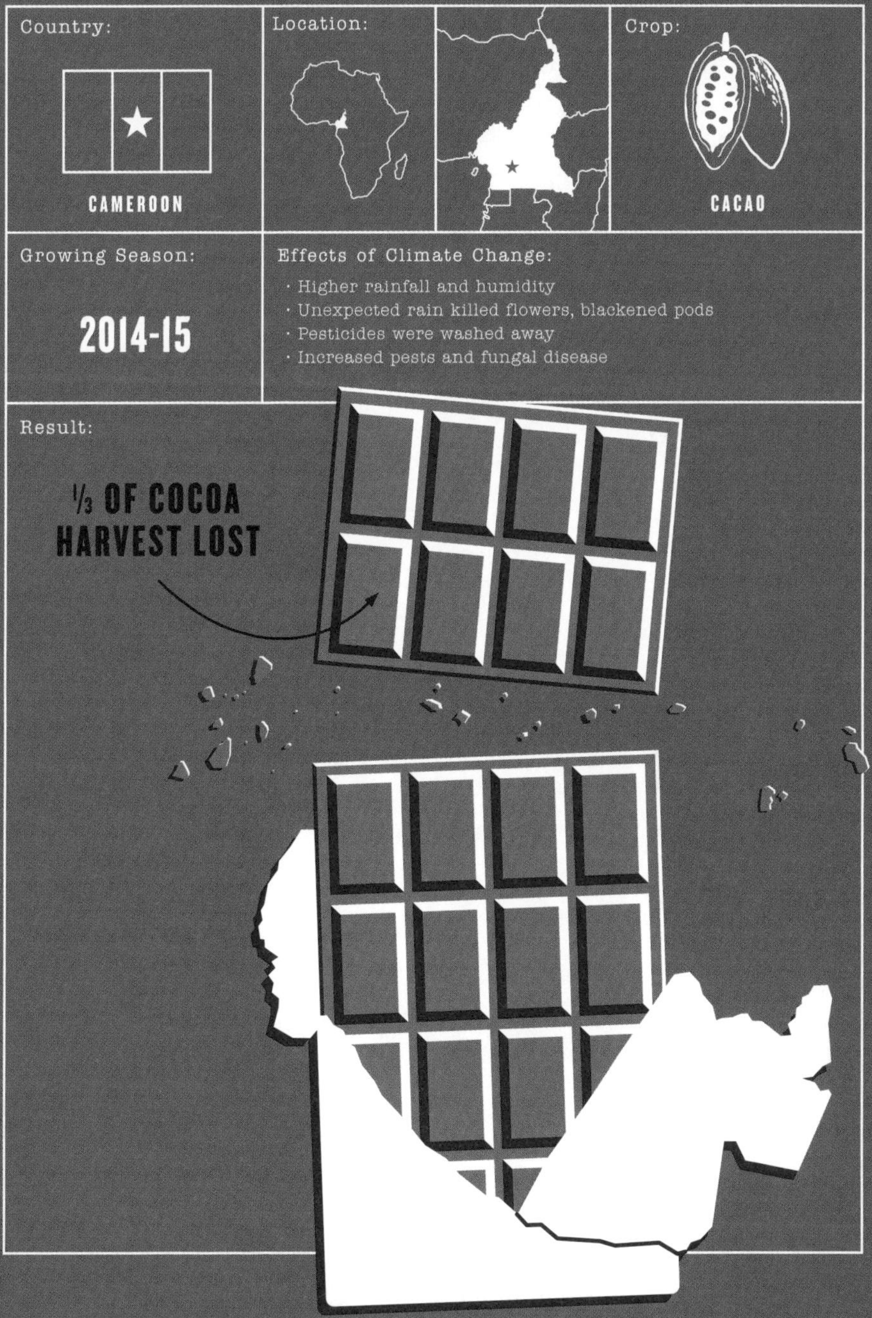

Country:
CAMEROON
Location:
Crop:
CACAO
Growing Season:
2014-15
Effects of Climate Change:
· Higher rainfall and humidity
· Unexpected rain killed flowers, blackened pods
· Pesticides were washed away
· Increased pests and fungal disease
Result:
⅓ OF COCOA HARVEST LOST

EFFECTS OF CLIMATE CHANGE ON THE WINE INDUSTRY

TOP TO BOTTOM, LEFT TO RIGHT: 1. Grown at higher elevations 2. Grown increasingly farther from equator 3. Wildfire 4. Declining yields and quality 5. Shifting growing regions 6. Earlier harves 7. Groundwater salinization 8. Smoke 9. Quicker ripening 10. Fungi, mildew 11. 50% reduction premium grape-growing areas 12. Negative impacts on outdoor workers 13. Soil erosion 14. Extreme heat 15. Pests 16. Rising sea levels and flooding 17. Soil desertification 18. Early buds succumb to frost 19. Excessive rain

ACTION 37

IMBIBE WITH CLIMATE IN MIND

Climate change is coming for our cocktails, beer, and wine! Shifting growing regions, more pests and disease, more extreme heat, sea-level rise, and wildfire and wildfire smoke will alter and stress the beer, wine, and spirits industries. For an industry that has tens of billions of dollars of economic impact in the United States each year, climate change is a real problem. For example, research shows that extreme heat could lead to a 50 percent reduction in the area where premium wine grapes can be grown. Globally, suitable regions for viticulture are expected to decrease over the next 30 years by between 19 to 62 percent with moderate emissions of greenhouse gases, or 25 to 73 percent with high emissions of greenhouse gases.

Drought, heat, wildfire, and water stress can impact the key ingredients in beer and alter the taste of wine. They also bring negative impacts to the outdoor workers that sustain the industry. Sea-level rise could be bad for some sugarcane-growing regions, spelling trouble for your favorite rum. And the list goes on!

What can you do to drink more responsibly when it comes to climate and the environment? While you might have to pay more to imbibe, you can also support breweries, wineries, and distilleries that are working to adopt more sustainable practices and that are advocating for more climate action by governments. New Belgium Brewing has a comprehensive climate change program, which spans from clean energy to packaging, and has been ranked number one by Better World Shopper for its leadership in social and environmental responsibility. There are currently over 200 US brewers and over 120 US wineries that have signed a declaration in support of climate action and carbon pricing. There is even a carbon footprint calculator for brewers. Cheers to climate action by industries and consumers!

▸ **What's in your pint glass or wineglass? What climate-friendly practices has your favorite winery, brewery, or distillery adopted?**

ACTION 38

GET TO KNOW YOUR FAVORITE COFFEE

The staple of most of our morning and afternoon routines is a nice cup of coffee. Stats back this up: in 2018, global consumption of coffee was close to 22 billion pounds, and demand is expected to triple by the year 2050. Climate change might have other plans for our daily coffee habit. As the planet warms, changing weather extremes—including more frequent droughts and heat waves—are expected to have a significant negative impact on coffee growers and the supply of coffee beans.

The global north consumes the most coffee, with the United States topping the list, yet most coffee is grown in equatorial and tropical climates. This distance between producer and consumer means that about 15 percent of the emissions associated with the production and consumption of coffee is associated with transportation. Further complicating matters is that a majority of us rely on coffee from only two key species—Coffea arabica, known as Arabica, and Coffea canephora, known as Robusta—with Arabica the species most widely used by large companies like Starbucks.

This limited diversity of species means that the coffee of today is unlikely to be the coffee of the future. As the climate warms, it is expected that not only will the coffee supply chain be stressed, due to more challenging growing conditions, but that there will also be a shift to more Robusta beans, as these trees can withstand hotter temperatures and less shade. Generally deemed a lower-quality coffee, many companies and nonprofits are investing in more climate-resistant coffee varieties, restoration of agricultural lands, improved land management, and farmer education, in hopes of better supporting the many small landholders and farmers who produce a bulk of the global coffee supply.

As consumers, we can buy shade-grown and organic coffee, support companies that are investing in climate adaptation practices, and support the farmers whose livelihoods are connected to this global industry. We can also avoid letting that cup of joe get tepid and go to waste (reheat it, or, hello, iced coffee!). We should also steer clear of disposable cups and pods, which come with their own climate impacts.

▸ **What's in your cup of joe?**

COFFEE: WHO MAKES IT AND WHO TAKES IT?

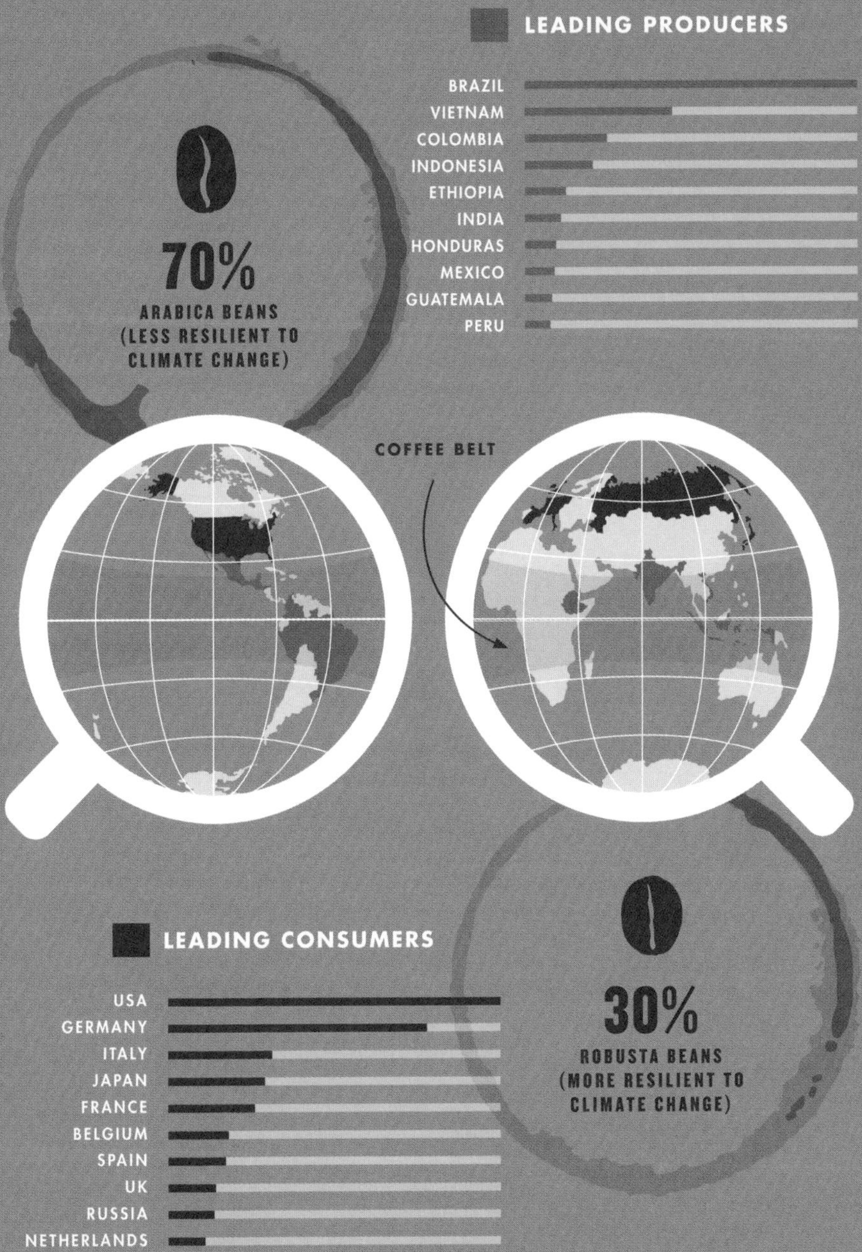

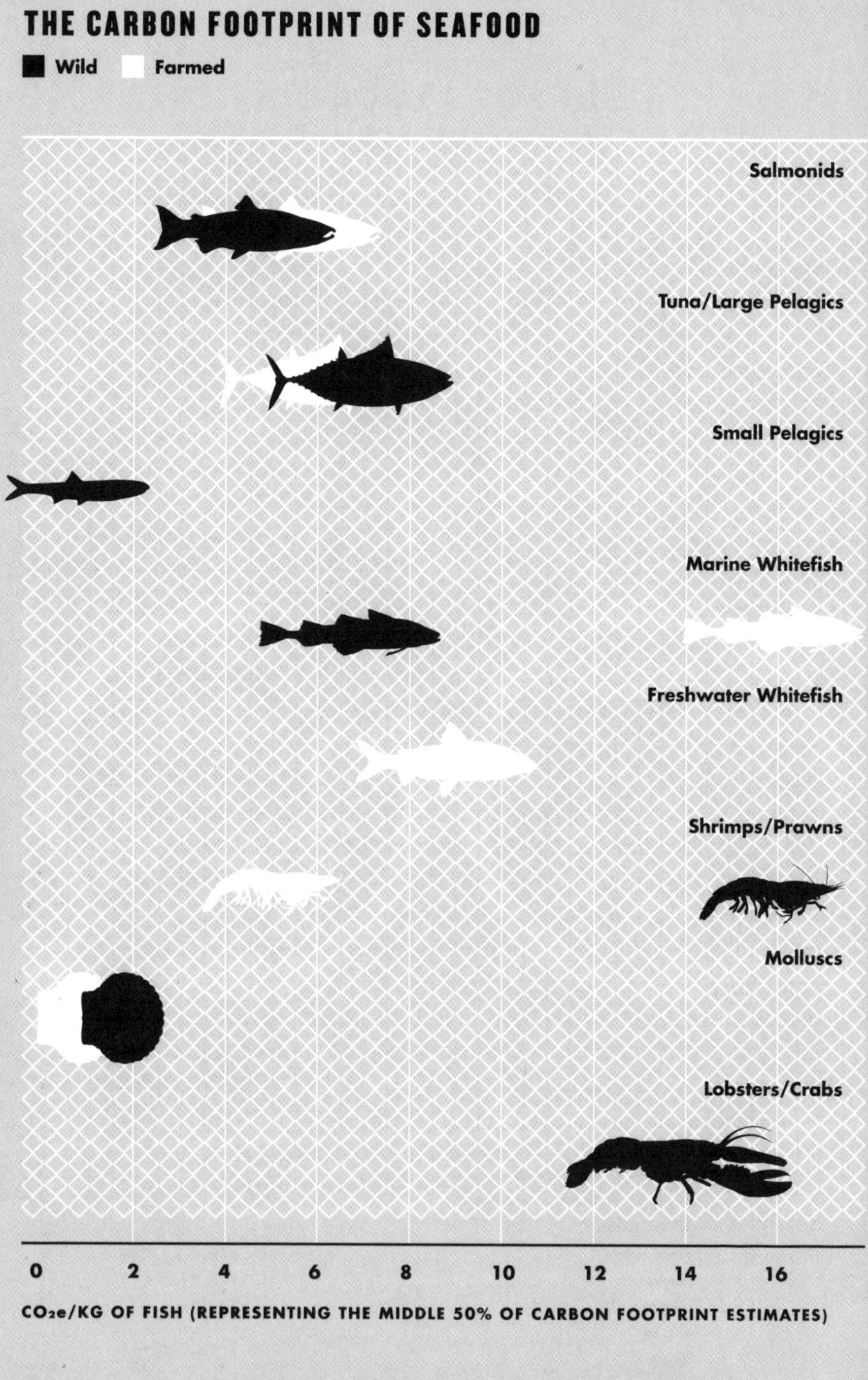

IN 2011, EMISSIONS FROM GLOBAL FISHERIES WERE ESTIMATED AT 179 MILLION METRIC TONS OF CARBON DIOXIDE EQUIVALENTS PER YEAR.

ACTION 39

SUPPORT LOCAL, SUSTAINABLE FISHERIES

Climate change is increasing ocean water temperature and acidity, causing sea levels to rise and impacting the health and function of coastal ecosystems and the communities that rely on the ocean and its wealth of cultural and natural resources. Did you know that the ocean has absorbed more than 90 percent of the excess heat in the climate system created by human-caused global warming? This is leading to impacts on the abundance and health of global fish and shellfish stocks and is expected to challenge global food supplies, tourism, and the health of communities with high levels of seafood consumption, including across the Arctic, West Africa and many island nations. Under high emissions of greenhouse gases in the future, the maximum catch potential of global fisheries could decline by 24 percent by 2100. This is very likely an impact three to four times higher than with significant reductions in emissions.

In 2011, emissions from global fisheries were 179 million metric tons of carbon dioxide equivalent greenhouse gas emissions per year, the same as the emissions from more than 2.3 million tanker trucks' worth of gasoline. For the US fishing industry, this translates to about 3.5 pounds of carbon emissions for each 2.2 pounds of fish caught. Sardines, anchovies, and shellfish tend to be good choices for their high protein content and lower carbon footprint. Crab, shrimp, and lobster tend to have a comparatively larger carbon footprint. Wild-caught fish can come with a smaller carbon and environmental footprint than farmed fish, while processed fish can result in around 8 percent less waste than fresh fish! Processed or frozen fish also has the advantage of having a smaller transportation-based emissions footprint, as most of these products travel by boat rather than a plane, like fresh fish.

As a consumer, consider sourcing fish as locally as possible. Check out SeafoodWatch.org, which provides a comprehensive database of the different impacts of seafood choices, based on harvest method, management practices, region, and other variables.

▸ **What fish dishes will be on your table this week?**

SHOPPING AND CONSUMER CHOICES

ACTION 40

REDUCE CONSUMPTION THROUGH COMMUNITY SHARING

Many of us are familiar with the "reduce, reuse, recycle" adage, but is there any climate merit to such actions? As it turns out, when we add "refurbish and repurpose" to the list, there are even more climate benefits. Repairing, refurbishing, and remanufacturing, or making new products out of existing ones that have run their useful life span, can reduce greenhouse gas emissions from some manufacturing sectors by anywhere from 79 to 99 percent! This process of shifting toward reduction and reuse, often called a circular economy—rather than making a product and then disposing of it—can bring clear climate and environmental benefits.

Since most of us are not directly engaged in manufacturing, how can we make an impact? We can use our purchasing power to buy goods that are refurbished or remanufactured and work to reduce our overall consumption of products, like clothes and shoes (see Action 41) and technology equipment (see Action 48). On a smaller scale, we can look to places where we can share items rather than purchase a product for a single use. Take tools, for example. Consider participating in a local tool library, where you can share tools and resources with your neighbors rather than fill your toolbox (while emptying your wallet!) with a specialized tool that might get only one or two uses. The Town of Telluride in Colorado manages the "Telluride Free Box" where community members drop off household goods, books, and other treasures looking for a new home. Tool libraries, free boxes, and Little Free Libraries are all examples of opportunities to share, repurpose, and give new life to items you no longer want. Want another adage? Sharing is caring . . . for your community and the climate!

▸ **What tool, book, or other item might you share with a neighbor?**

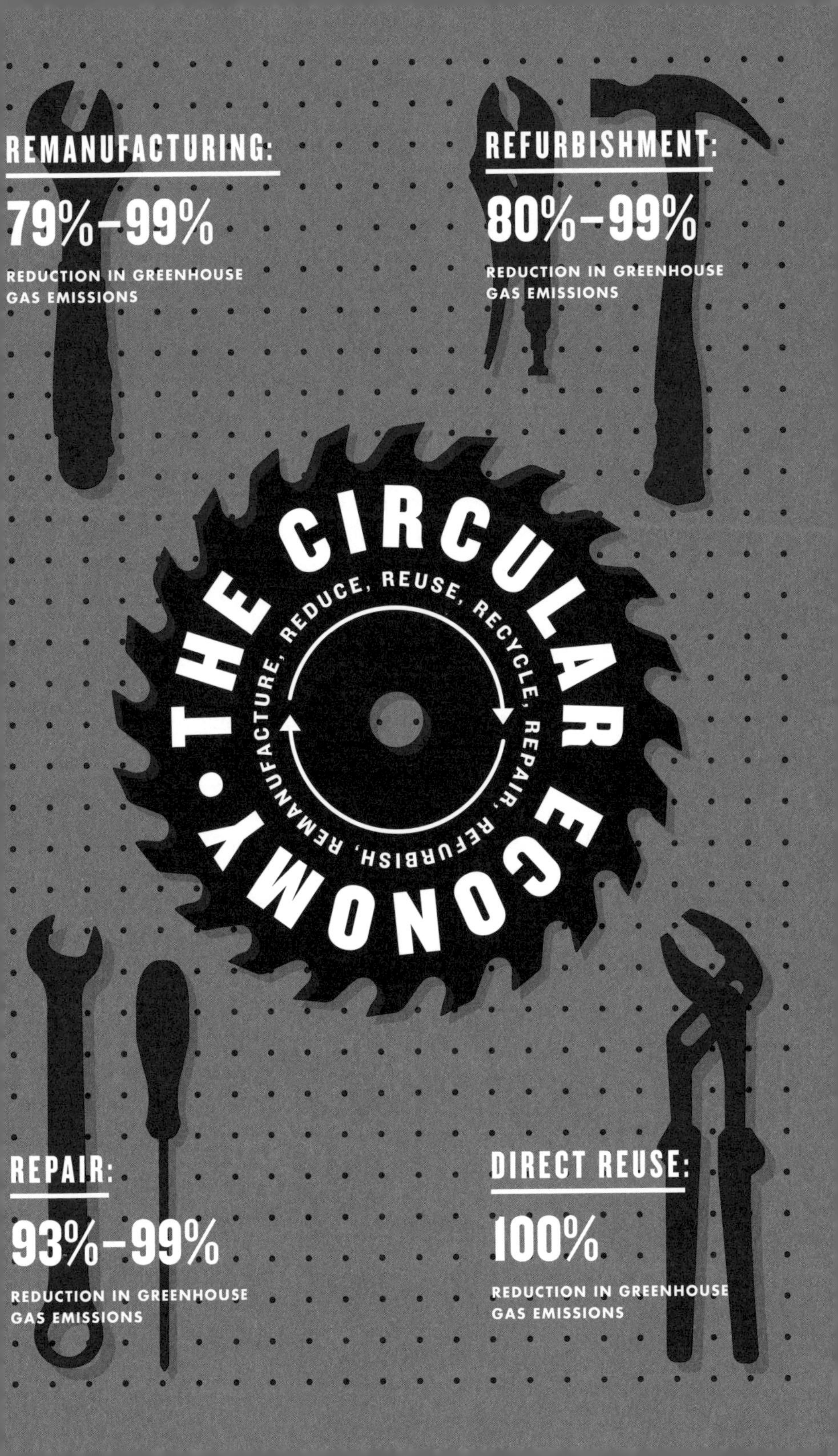
REMANUFACTURING:
79%–99%
REDUCTION IN GREENHOUSE GAS EMISSIONS
REFURBISHMENT:
80%–99%
REDUCTION IN GREENHOUSE GAS EMISSIONS
THE CIRCULAR ECONOMY
REDUCE, REUSE, RECYCLE, REPAIR, REFURBISH, REMANUFACTURE,
REPAIR:
93%–99%
REDUCTION IN GREENHOUSE GAS EMISSIONS
DIRECT REUSE:
100%
REDUCTION IN GREENHOUSE GAS EMISSIONS

ACTION 41

TURN AWAY FROM FAST FASHION

The clothes we wear have a clear climate connection—from emissions produced during production and transport to waste from manufacturing, and water consumption and chemicals used for crops like cotton that wind up in our T-shirts and jeans. Estimates put waste from the fashion industry at over 92 million tons per year! Further, clothing with a short life span, typified by the concept of "fast fashion," often ends up in a landfill or incinerated. In fact, 73 percent of discarded textile materials end up buried or burned, with only 1 percent being recycled into new clothing.

This wasteful trend—alongside the reality that the average number of times a garment is worn has decreased by 36 percent over the last 15 years while global clothing production has nearly doubled—means we are producing more clothes, more waste, and more emissions. The apparel and footwear industry account for 6 to 10 percent of global greenhouse gas emissions. While many companies are coming up with greener ways for us to wear their looks, a green label doesn't necessarily mean a garment is climate friendly. New business approaches, including clothing rentals and recycling, often come with a higher climate price tag than embracing the first two Rs in "reduce, reuse, recycle."

To reduce the climate cost of your closet, buy fewer items, wear your clothes and shoes longer, and pass along or sell your unused or rarely used favorite looks. When you need to refresh your look, buy used or invest in high-quality products and staple pieces that can be mixed, matched, and worn for years. You'll not only look good but you will lower your carbon footprint in the process.

▸ **What's in your closet?**

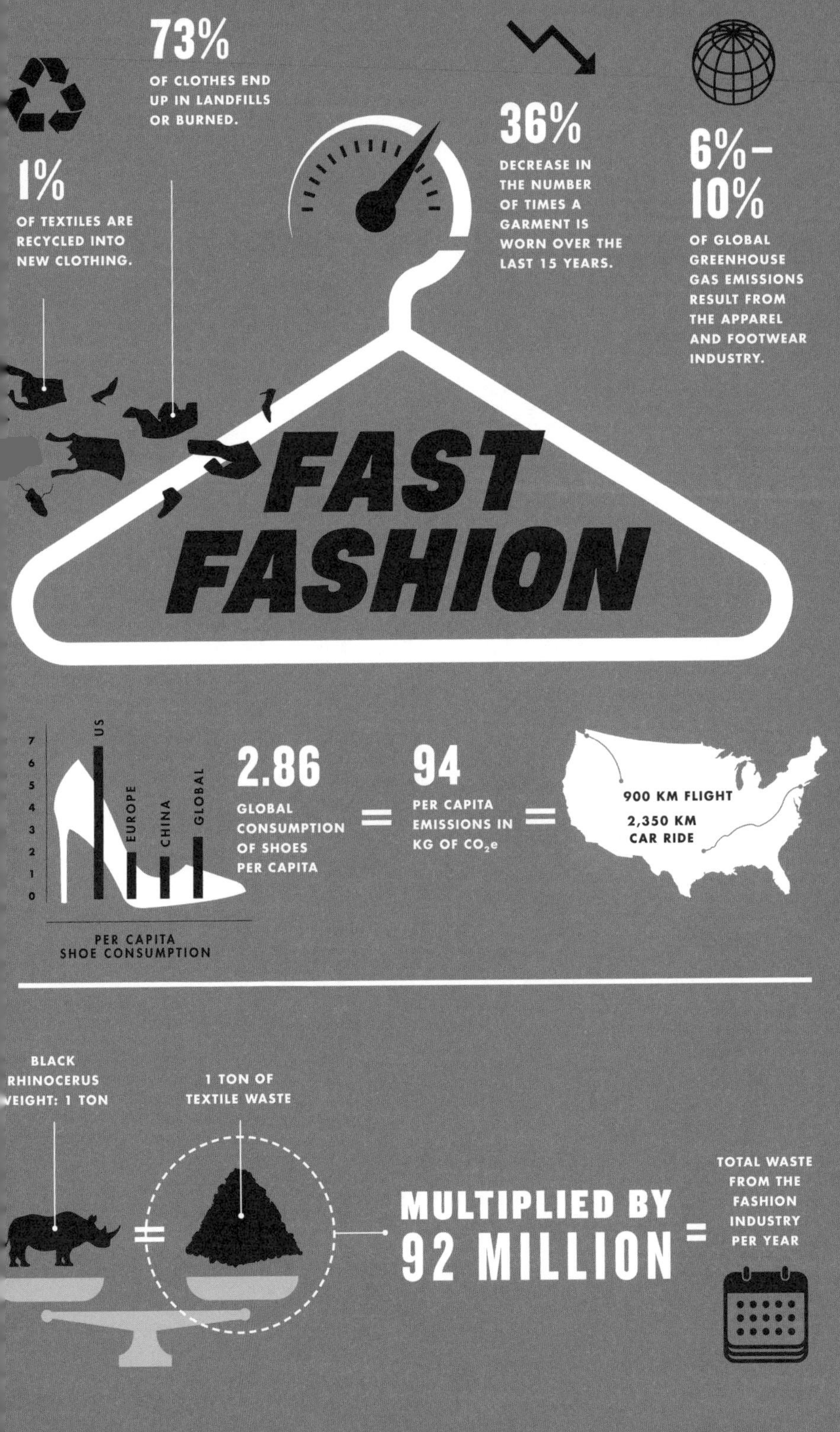

73%
OF CLOTHES END UP IN LANDFILLS OR BURNED.
1%
OF TEXTILES ARE RECYCLED INTO NEW CLOTHING.
36%
DECREASE IN THE NUMBER OF TIMES A GARMENT IS WORN OVER THE LAST 15 YEARS.
6%–10%
OF GLOBAL GREENHOUSE GAS EMISSIONS RESULT FROM THE APPAREL AND FOOTWEAR INDUSTRY.
FAST FASHION
7
6
5
4
3
2
1
0
US
EUROPE
CHINA
GLOBAL
PER CAPITA SHOE CONSUMPTION
2.86
GLOBAL CONSUMPTION OF SHOES PER CAPITA
=
94
PER CAPITA EMISSIONS IN KG OF CO_2e
=
900 KM FLIGHT
2,350 KM CAR RIDE
BLACK RHINOCERUS WEIGHT: 1 TON
1 TON OF TEXTILE WASTE
MULTIPLIED BY 92 MILLION
=
TOTAL WASTE FROM THE FASHION INDUSTRY PER YEAR

95% OF BABIES IN THE UNITED STATES WEAR DISPOSABLE DIAPERS.	5% OF BABIES IN THE UNITED STATES WEAR CLOTH DIAPERS.
133 DISPOSABLE DIAPERS	= 1 CLOTH DIAPER
550 KILOGRAMS CARBON DIOXIDE EQUIVALENT EMISSIONS FROM PRODUCTION TO TRANSPORTATION	570 KILOGRAMS CARBON DIOXIDE EQUIVALENT EMISSIONS FROM PRODUCTION TO TRANSPORTATION
WHICH IS EQUIVALENT TO 62 GALLONS OF GASOLINE	WHICH IS EQUIVALENT TO 64 GALLONS OF GASOLINE
141 CUBIC METERS OF WATER USED	1,221 CUBIC METERS OF WATER USED
MADE FROM PETROLEUM PRODUCTS	MADE FROM COTTON, WHICH CAN USE LARGE AMOUNTS OF WATER AND PESTICIDES

ACTION 42

WEIGH YOUR DIAPER OPTIONS

Many parents try to limit the environmental impact of their children. This is no surprise, given that recent research suggests that one of the largest climate actions a family can take in developed countries is having one fewer child—saving over 58 metric tons of CO_2e. While reducing the number of people on the planet who are drawing from a finite pool of resources is a critical factor in climate change and environmental sustainability, family size is a deeply personal choice. For those of us with kids and growing families, how can we limit the impact of our families on the planet?

Climate and environment considerations start at birth, with decisions around goods like diapers, clothing, and gear. Most parents are faced with the question of disposable or cloth diapers. Which is better for the environment? Diapers dominate the first 2.5 years of most kids' lives, with 95 percent of babies in the United States wearing disposable diapers, producing about 7 percent of nondurable household waste in landfills each year. A life-cycle analysis of diapers, from emissions and material production to transportation, puts the carbon footprint of disposable diapers over the first 2.5 years of a child's life equivalent to the emissions from burning 62 gallons of gasoline.

Cloth diapers, while reusable, come with the climate cost of cotton, which often involves water and fertilizer, and the added energy associated with laundering. The amount of water used, the temperature of the water being used for washing, and the type of electricity used to dry reusable diapers all contribute to the climate-cost calculus. Emissions from cloth diapering over the first 2.5 years of a child's life is about 570 kilograms of CO_2e—or the equivalent of burning about 64 gallons of gasoline. The emissions numbers between cloth and disposable diapers are close, so be sure to consider the climate context where you live. (Do you live in a drought stressed area? Opting for disposable might be a better choice.) Fortunately, many popular brands are limiting the use of petroleum-derived ingredients, and are working to lower the weight and bulk of diapers to reduce transportation-related emissions and to occupy less space in the landfill. Finding a brand on the move toward more sustainable diapering offers you an opportunity for managing the mess with the environment in mind.

▶ **What diapers will you choose?**

ACTION 43

DITCH THE BOTTLED WATER

Taste, perceptions of safety, convenience, and effective marketing have all driven an increase in the consumption of bottled water. In 2017, nearly 100 billion gallons of water were bottled globally, with 13.2 billion gallons produced in the United States alone. That number rose to 13.8 billion gallons in 2018, and industry reports suggest real bumps in consumption during the COVID-19 pandemic. This eye-popping amount of water comes at a real environmental cost, but it also drives a multibillion-dollar industry. Recent profits were reported at just under $20 billion in the US market across all bottled-water categories.

There are several nuances that determine the total environmental and climate impact of a given bottle of water, including the water's original source, the energy used to extract and treat the water, the type and weight of the packaging, the volume of water in the container, the distance between the source and the consumer (e.g. If you live in the United States you might want to reconsider bottled water from overseas), and how the bottle is disposed of. Many studies have unpacked these nuances, and a recent one put it succinctly: "Tap water always has the best environmental performance, even in case of high energy-consuming technologies for drinking water treatments."

To put some numbers on it, it takes about 12 cups of water to produce about 4 cups of bottled water. It is also estimated to take about 1,000 to 2,000 times more energy to produce bottled water than tap water. While lack of potable water access can necessitate bottled water use, for those who are fortunate enough to have access, it is better for the environment to opt for tap water. If you don't like the taste or are worried about the quality of your drinking water, ask your water utility or supplier for its water quality information. You can test your tap water or well water with kits that are mailed to your home. You can also opt for an at-home filtration system that attaches to your sink, or a refillable jug for your fridge.

▸ **Will you reduce your consumption of bottled water?**

IT TAKES

1,000 TO 2,000

TIMES MORE

ENERGY TO

PRODUCE BOTTLED

WATER THAN

TAP WATER.

ACTION 44

CUT DOWN ON PLASTIC

At the heart of plastics is fossil fuel. Petrochemicals, oil, natural gas, and coal are part of the creation and distribution of plastics. The life cycle of plastic—from the chemical building blocks and its manufacturing to its disposal—results in a significantly large carbon footprint. Today, the plastic life cycle results in emissions equivalent to those produced by 189 coal plants each year. Most plastic also sticks around, taking hundreds of years to decompose. Looking to the future, without changes in the patterns of consumption and (re)use of plastic, emissions could skyrocket to the equivalent of what is produced by 615 coal plants, or about 13 percent of our total global carbon budget!

As we seek solutions, we need to consider the many ways plastic appears in our lives. Globally, packaging is a primary source of plastic in a regular consumer's life. In fact, around 40 percent of plastics produced are utilized in packaging. This packaging is often single use, leading to near-immediate disposal. While some of our plastic packaging is recycled, much of this waste ends up in landfills or is burned.

Plastic has positively contributed to society, but there are numerous downsides besides climate costs—including human health and environmental justice dimensions—to our obsession with, and rapid disposal of, plastic. For example, plastic burning often occurs near low-income communities and communities of color, and is connected to chemical exposure and reduced air quality.

It is vital that we all explore ways to reduce our plastic consumption. Start with an audit of plastic in your home. You might be surprised by all of the places plastics appear in your life—from your shower to your suit coat. There are also a growing number of options to recycle and repurpose traditionally hard-to-recycle products, like plastic film, through companies like Ridwell. We look at our daily habits, it is critical to encourage the manufacturing industry to accelerate the reduction, recycling, reuse, and remanufacturing of plastic and invest in innovations to reduce the climate and environmental impacts of plastic.

▸ **Will you opt for less plastic in your life?**

AT THE HEART OF PLASTICS IS FOSSIL FUEL.

Around 40% of plastics produced are utilized in packaging and often single use, leading to near-immediate disposal.

ACTION 45

AVOID MICROPLASTICS

Tiny fragments and threads of plastic—from our synthetic clothing, personal care products, tires, disposable containers, and larger plastic products—are called microplastics. Microplastics are technically the size of sesame seed (5 millimeters) or smaller and have been documented in just about every corner of the globe—across our oceans, inside our bodies, in the food we eat and the water we drink, in wildlife and shellfish, in the snow in our mountains, and even in Antarctica!

Beyond the emissions of greenhouse gases associated with the production, transit, and breakdown of plastic (see Action 44), microplastics have interesting climate connections. Microplastics in the ocean are expected to lead to declining ocean health, particularly because these fine particles can be consumed by zooplankton. Zooplankton are critical in the marine food chain. If they eat microplastics in lieu of natural food, it can lead to more algal blooms, reduce the amount of oxygen in the water, and negatively influence the overall health of the ocean. It is also hypothesized that microplastics in snow, in addition to particles produced from wildfire smoke and from combustion of fossil fuels, may help accelerate the melting of snow because darker surfaces absorb more energy from the sun accelerating the melting of snow. This can negatively impact rivers, ecosystems, and the water supplies many communities rely on.

You can help limit the production of microplastics by reducing your overall plastic consumption, avoiding personal care products that contain microbeads and plastics (often noted by the presence of copolymers on the ingredient list), and avoiding synthetic clothing. While the United States banned microbeads in rinse-off cosmetics and toothpastes in 2015, plastics can still sneak into our personal care products (see Action 78). Also consider buying a fine mesh bag, sold at retailers like Patagonia, for washing your synthetic clothes—like fleece jackets, which are notorious for producing microplastic threads. These bags trap microplastics, which can then be thrown away in your trash, rather than sent down the drain and into our waterways, ecosystems, and oceans.

▸ **Will you commit to using and producing fewer microplastics?**

1900 MICROFIBERS ARE RELEASED FROM EACH GARMENT IN A LOAD OF LAUNDRY

SOURCES OF MICROPLASTICS:

- SYNTHETIC TEXTILES
- TIRES
- CITY DUST
- PERSONAL CARE PRODUCTS
- MARINE COATINGS
- ROAD MARKINGS
- PLASTIC PELLETS
- BREAKDOWN OF LARGER PLASTICS

35%
OF MICROPLASTICS IN THE OCEAN COME FROM SYNTHETIC TEXTILES

80%
OF ALL MICROPLASTICS COME FROM TEXTILES, TIRES, AND CITY DUST

WHERE MICROPLASTICS ARE FOUND:

- OCEAN
- SEA ICE
- FRESHWATER
- SOIL
- HUMAN BODY
- WILDLIFE
- AIR

MICROPLASTICS

ACTUAL SIZE

5 MM OR LESS

11
PIECES OF MICROPLASTIC WERE FOUND ON AVERAGE IN EACH SPECIMEN.

2
OYSTERS DID NOT.

298
OYSTERS CONTAINED MICROPLASTICS FROM TEXTILES.

OREGON

300 OYSTERS AND RAZOR CLAMS SAMPLED FROM THE OREGON COAST WERE TESTED FOR MICROPLASTIC CONTAMINATION.

fast shipping **Warehouse Facilities are**

What are you looking for?

Often Close to Low-Income and Minority Communities

E-commerce and the Environment

★☆☆☆☆ (165,000,000,000 packages per year | 500 per person)

8.8% of total retail sales in US

$448.3 billion in sales

Year: 2017

& FREE Returns

For **fastest delivery** see information on following page

Select delivery location

In Stock

Qty: 1

Add to Cart

Buy Now

Secure transaction

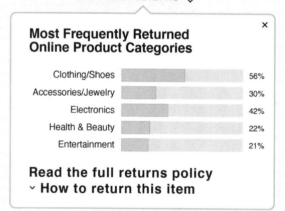

Most Frequently Returned Online Product Categories

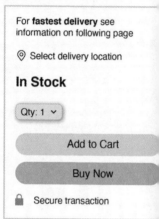

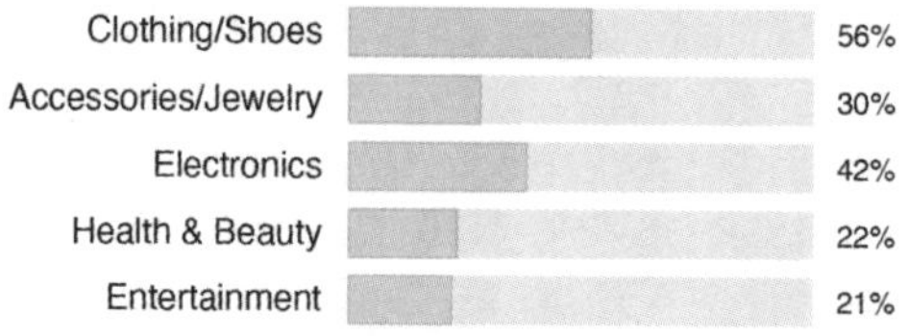

Read the full returns policy

How to return this item

Product Details

- **30% of shoppers "overbuy" and end up returning products.**
- **Delivery trucks tend to produce around ten times more nitrous oxide (NOx) compared to passenger cars.**
- **E-commerce has increased NOx in some neighborhoods by as much as 24%.**

ACTION 46

MAKE MORE THOUGHTFUL ONLINE PURCHASES

The COVID-19 pandemic further cemented the trend toward online shopping, which has been on an upward trajectory over the last decade. In 2020, just 10 large retailers were responsible for 68 percent of online shopping sales in the United States. Amazon alone accounts for more than 50 percent of all online sales. As many of us click and buy our way around the internet and have packages delivered, sometimes just hours later, many shoppers wonder about the environmental and social impacts of our online purchases. Buckle up for some staggering statistics: It has been estimated that 165 billion packages are shipped in the United States each year. That's about 500 packages per person! E-commerce represented almost 9 percent of all US retail sales in 2017.

To be clear, e-commerce isn't necessarily worse for the environment when compared to brick-and-mortar shopping; the consolidation of many packages into one truck on a well-coordinated and efficient route around town can actually reduce the number of us traveling in our own cars to stores. One study likened this process to mass transit. Just as putting a bunch of people on a bus going in the same direction on a well-planned route leads to greater efficiency and less emissions, so, too, can a coordinated and well-planned shipping and delivery route. Improved coordination and efficiency could reduce up to 84 percent of emissions from e-commerce transportation!

E-commerce requires a significant amount of warehouse space, and the current locations of warehouses are often close to minority and low-income communities, disproportionately exposing them to the pollution, noise, and traffic created by these facilities. Delivery trucks tend to produce around 10 times more nitrous oxide than passenger cars, and e-commerce has increased nitrous oxide in some neighborhoods by as much as 24 percent. Online shopping also results in more returned products, increasing the associated emissions of our online purchases. In fact, 30 percent of shoppers "overbuy" and end up returning products. The return rate for clothes and shoes is a staggering 56 percent!

▸ **Will you commit to shopping online with intention and consolidating purchases when you can?**

ACTION 47

SLOW DOWN YOUR SHIPPING

While online shopping can, in some cases, be more environmentally friendly, the major climate cost of e-commerce often comes with the speed of shipping. The increasingly popular and available one- and two-day—or even one-hour—delivery options come with a real impact. As it turns out, there is a clear correlation between the speed of delivery and the emissions associated with a given purchase. Put another way, speedy shipping may be convenient, but it spells trouble for the environmental impact of our online shopping. One 2020 study measured a 180 percent increase in emissions associated with the shortening of delivery windows.

These shorter windows limit the options companies have for consolidating orders to maximize efficiency of their deliveries and reduce the number of miles traveled for each order. The final distance between the product and the consumer is often referred to as the last mile. This last mile can pack a real punch when it comes to emissions and environmental impact. The difference in vehicle miles traveled for a package can increase by nearly six times when a one-hour delivery window is opted for over a 24-hour delivery window.

In addition to speed, the products we buy and the companies we buy them from also influence the overall social and environmental impact of our purchasing. So what can we do?

1. Use recurring deliveries, sometimes known as "auto-ship," for staple products that you regularly buy, to help vendors anticipate demand and optimize delivery of those products to your doorstep.
2. Opt for slower and combined shipping.
3. Buy only what you need, and avoid returns. If in doubt about sizes, shop for clothes and shoes in person. However, resist the temptation to drive to a store to see or try on a product, only to turn around and buy it online.
4. Do your research on the companies you are purchasing from and try to shop at companies that share your values.

With a little planning and forethought, we can save money and reduce the climate impact of our online purchases.

▸ **Will you choose slower shipping and longer delivery windows?**

fast shipping

F

A 2020 study measured a **180%** increase in emissions associated with the shortening of delivery time windows.

FAST SHIPPING

SHIP TO: "The responsibility for improving last-mile sustainability is shared, beginning with the online shopper in consolidating their purchases, with the vendors and last-mile carriers in consolidating as many customers as possible into a single delivery tour, and with planners, regulators, and civic society in demanding/implementing improved tailpipe emission truck technologies."

Emissions per Package

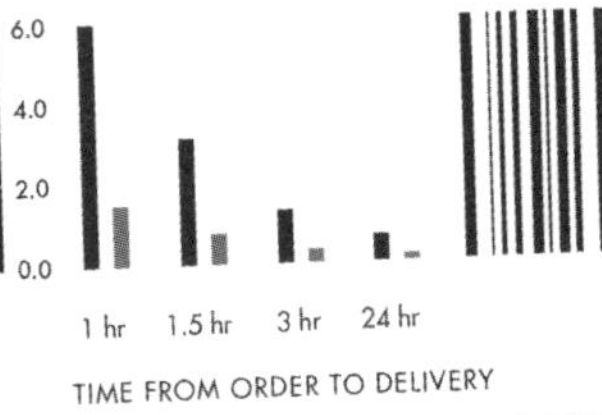

SPEED

EMISSIONS

fast shipping

fast shipping

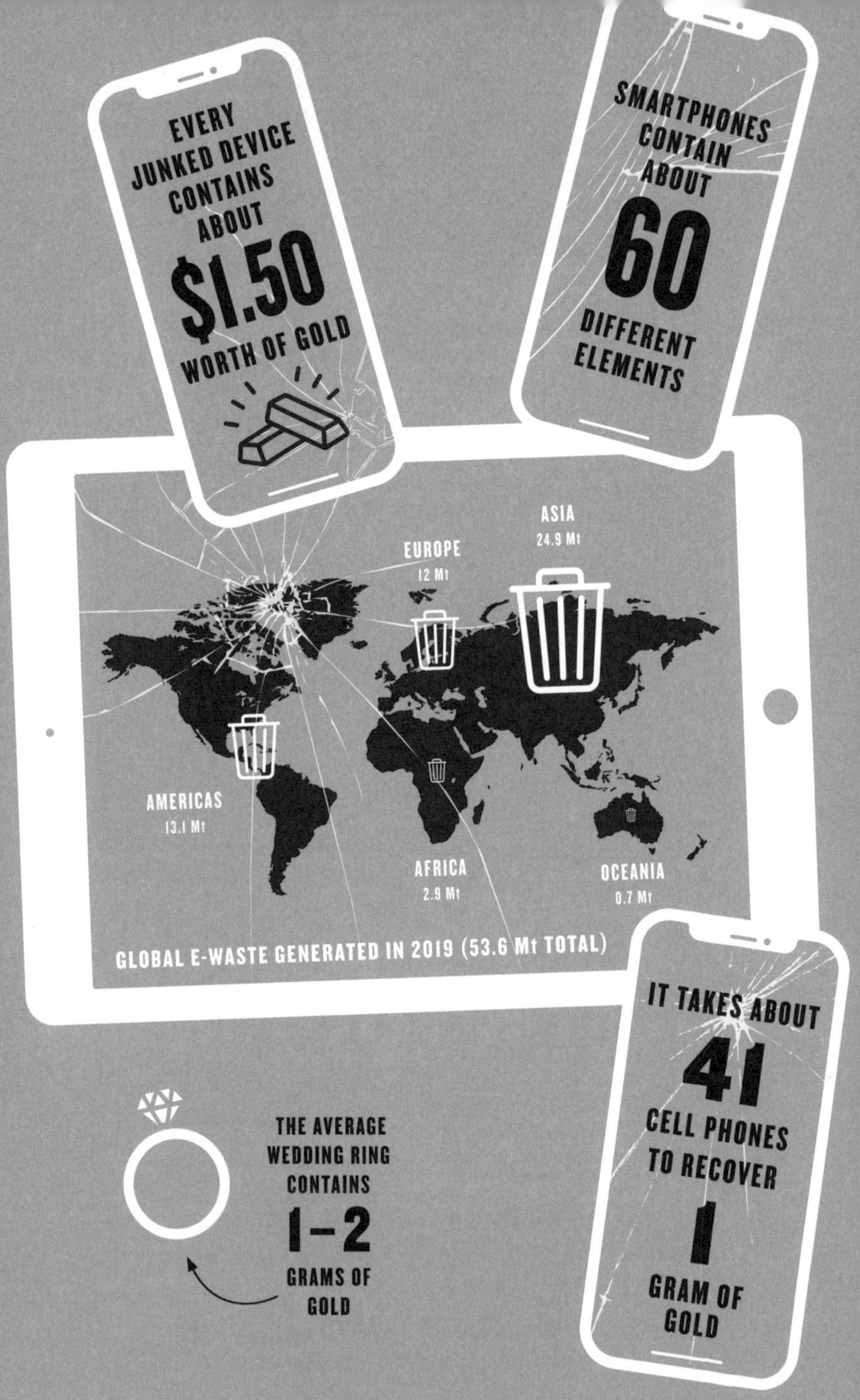
EVERY JUNKED DEVICE CONTAINS ABOUT
$1.50
WORTH OF GOLD
SMARTPHONES CONTAIN ABOUT
60
DIFFERENT ELEMENTS
ASIA
24.9 Mt
EUROPE
12 Mt
AMERICAS
13.1 Mt
AFRICA
2.9 Mt
OCEANIA
0.7 Mt
GLOBAL E-WASTE GENERATED IN 2019 (53.6 Mt TOTAL)
IT TAKES ABOUT
41
CELL PHONES TO RECOVER
1
GRAM OF GOLD
THE AVERAGE WEDDING RING CONTAINS
1–2
GRAMS OF GOLD

E-WASTE IS THE WORLD'S FASTEST-GROWING DOMESTIC WASTE STREAM

ACTION 48

KEEP YOUR DEVICES LONGER AND DISPOSE OF THEM PROPERLY

Open the junk drawer or favorite hiding spot for your old phones and other electronic devices. Did you know that both the devices collecting dust and the ones we are actively using come with a real climate impact? It isn't just the devices themselves—every text, email, funny meme, or online purchase we make requires a server and data center, which are incredibly energy intensive.

Smartphones themselves have a surprising impact. Emissions from smartphone manufacturing and use is estimated to be as much as 125 million metric tons of CO_2e per year. Around 85 to 95 percent of these emissions are produced during manufacturing. When we look at the life cycle of a smartphone—from mining and manufacturing to how it is disposed of—it adds up. The amount of waste produced in the European Union from manufacturing new electronics was five times greater than the amount of e-waste created by consumers disposing of old devices.

All of this waste is problematic, whether in the manufacturing or end-of-life stage. Electronic waste, a.k.a. e-waste—which refers to products at the end of their useful life—is a growing problem. It has been called a national security risk, is bad for our health, has real financial costs, and is an environmental justice issue. Globally, 53.6 million metric tons of e-waste was created in 2019, with an estimated $57 billion of reusable, high-value materials like gold, copper, and platinum being tossed or burned as a result of poor practices by consumers and corporations.

What can we do? Keep your electronic devices as long as possible, and avoid the urge to upgrade your phone every time a new version is available. Rather than replacing our phones every two years, keeping these devices, that require phenomenal resources to create, will help limit our impact. When you need to get a new phone, tablet, computer, or monitor, please properly recycle your old devices. Today, 99 percent of smartphones are not recycled. On a bigger scale, we also need to encourage policies and companies to get our data centers running on renewable energy.

▸ **Will you commit to keeping your devices as long as possible and creating less e-waste?**

ACTION 49

SHOP YOUR VALUES

Spending your money at companies that reflect your values helps keep these companies viable. Research also shows that shopping at places that reflect our values—whether they be social, environmental, political, or all of the above—can have a direct and positive impact on us. Despite the potential for inconvenience (say that values-aligned pet store doesn't provide one-hour shipping, but the big behemoth of a corporation does), more limited choice, or slightly increased cost, consumers who opt to shop in alignment with their values can have a greater sense of empowerment, social embeddedness, and fulfillment!

How do you know what the most ethical and climate-friendly shopping choices are? There's a platform for that! One of the tools I personally use when looking at different companies and where to spend my money is Better World Shopper (BetterWorldShopper.org). Created by sociologist Ellis Jones, this website is backed by peer-reviewed research and provides a grade for more than 2,000 companies based on a range of factors, including a company's environmental, social justice, human rights, and animal protection practices, and its community involvement. One way I've used this site is by looking for the best source for my dog Kuna's favorite food. After conducting my own research about carbon-neutral shipping options, the company that produces the food, and the online shop I use, I ended up purchasing Kuna's food from a retailer rated A+ by Better World Shopper. To my surprise, it isn't costing me any more than if I had chosen a retailer with a worse grade. Who doesn't want to get an A+?

I encourage you to also do your own research by looking at company materials, reports, and reviews. Be careful, though: greenwashing is becoming more common, which can make it even more difficult at first glance to understand which companies are really walking the walk (see Action 50).

▸ **Will you do some research to make sure your values are reflected by the companies where you spend your hard-earned money?**

WE CAN'T NECESSARILY TELL WHAT HAPPENS IN THE PATH FROM ONLINE ORDER TO OUR DOORSTEP. IF YOU'RE NOT SURE, MAYBE YOUR PURCHASE

CAN WAIT.

ACTION 50

BEWARE GREENWASHING

As someone whose search history involves lots of queries about all things related to climate change, I subsequently have news, social media, and web search pages plastered with ads extolling the virtues of many companies' sustainability efforts. A 2021 study by marketing consultants found that 34 percent of global consumers are willing to pay more for sustainable goods and services while 63 percent reported modifying their consumption behaviors to be more sustainable. These data make it clear there is a market for green and sustainable products. But are sustainability pledges, claims, and products good, or just too good to be true? Unfortunately, some companies have taken advantage of this green trend. When claims related to sustainability practices or sustainable products are in actuality a ruse or a slick marketing ploy, greenwashing is in play.

To be clear, not all companies make environmental claims that are false or nefarious. Many companies really are working toward more environmentally friendly practices, like removing toxic materials from their products, using more sustainable manufacturing processes, reducing packaging, designing products to be more easily repaired, and deploying recycling programs for when you are done using their products.

What can we do? Learn how to "call bull" on suspect information and marketing ploys you see pop up in your news and social media feeds (see Action 93). Unfortunately, it's hard for individuals to get the full picture when a company's purported values don't necessarily align with what it's lobbying for, because such activities are harder to track. Thus, many experts and consumers have called for greater transparency and corporate accountability.

While we push for these larger levers of action and accountability, we can also follow a few tips recommended by the news website EcoWatch, based on its review of the Federal Trade Commission's Green Guides, to avoid the traps of greenwashing:

- **IGNORE THE HYPE.** Look for packaging that explains a product's positive environmental impact in plain language and doesn't overstate or imply more benefits than could possibly be delivered.
- **CHECK FOR CLARITY.** Make sure you can tell whether the marketing claims refer to the product itself or just the packaging.
- **LOOK FOR CERTIFICATIONS.** If claims are being made, see if they are backed by trusted third-party certifications, like the Forest Stewardship Council's forest management certification, B Corp Certification, the US Department of Agriculture's organic certification, or fair trade certification.

▶ **Which companies' environmental claims are you going to research?**

GREENWASHING noun

green·wash·ing | \ ˈgrēn-ˌwȯ-shiŋ , -ˌwä- \:

expressions of environmentalist concerns especially as a cover for products, policies, or activities.

—Merriam-Webster.com

ACTIONS AROUND THE HOME

ACTION 51

CONSIDER WHERE YOU MAKE YOUR HOME

As we consider our exposure to the impacts of climate change, we need to be looking closely at the place we call home. Whether you rent or own, live on the coast or in the mountains, or somewhere in between, climate change is likely to both directly and indirectly impact you. In many parts of the United States, flood risk is increasing due to changing patterns in rainfall, intensifying hurricanes, and rising sea levels, while other areas are facing increased wildfire and drought-related risks. These impacts are leading to significant costs, damages, and loss of life. In 2021, there were 20 billion-dollar weather and climate disasters in the United States, causing $152.6 billion in damage.

Why does this matter to you, especially if you're lucky enough to not have directly experienced one of these events? Take, for example, the increased costs from these events and the growing risk of more properties and people being exposed to these impacts. The United States coastal area are home to 42 percent of the population. With rising sea levels, many people, property, critical habitat, and infrastructure like roads and ports are at risk. Critically, this is not going unnoticed by insurance companies, who are increasing the cost of homeowner's insurance for many of us, or dropping coverage altogether. The Federal Emergency Management Agency (FEMA) is also planning to reduce subsidies for flood insurance.

What can you do? Look at local, state, or federal risk maps to understand your climate risks. Concerned about floods, for example? Look at flood maps supplied by FEMA or your local municipality (and be to sure to ask if the maps account for future climate change). If you live on the coast, check out the National Oceanic and Atmospheric Administration's Coastal Flood Exposure Mapper. Wondering about wildfires? Check out the US Forest Service's Wildfire Risk to Communities interactive maps and resources. Understand your risks, advocate for solutions that protect you and your community, and ensure you are insured for the risks you might face (see Action 52).

▸ **What climate risks does your home and community face? What climate-related costs might you need to plan for in the future?**

IN 2021 THERE WERE

20 BILLION-DOLLAR

WEATHER AND CLIMATE DISASTERS IN THE UNITED STATES, CAUSING

$152.6 BILLION

IN DAMAGE.

—National Oceanic and Atmospheric Administration

ACTION 52

CHECK YOUR INSURANCE POLICY

Climate change is a hot topic in the insurance industry. Why? Because it's making it more difficult to make a profit as costs increase for paying out for damages related to extreme weather. According to the Swiss Re Group, one of the world's leading providers of reinsurance and insurance globally, "extreme weather events in 2021, including a deep winter freeze, floods, severe thunderstorms, heatwaves, and a major hurricane, resulted in annual insured losses from natural catastrophes estimated at $105 billion, the fourth highest since 1970."

As a result, many home and rental insurance premiums are going up, and in some cases, people are being denied insurance based on their exposure to increasing climate risks. As more "no frills" policies are made available, and different terms and conditions and deductibles are put in place for weather-related disasters—like for named storms (e.g., Hurricane Ida)—we may find we no longer have the protection we need.

Governments are also taking notice. In 2021, the Federal Emergency Management Agency updated its flood insurance rating methodology to better reflect current-day flood risks. In some cases, this update is expected to translate to increases in flood insurance premiums. Curious about the impact of this on costs in your community? Check out the interactive maps created by the Association of State Floodplain Managers, which show anticipated premium changes. Some people are also now at risk of losing their insurance, prompting some state insurance commissioners to put moratoriums on coverage cancellation.

While the government needs to make larger regulatory changes, individuals should make sure they are covered for the risks they might face. Closely check your insurance policy and consider what coverage you might need. Call your insurance company to understand your coverage, consider adding flood or fire protection, and make sure to check your rates every year. It is expected that the cost of premiums is going up for most of us. Who knew that one of our least favorite "adulting" tasks could count as climate action?

▶ **Will you check your insurance coverage with climate in mind? How can you prepare for premium increases?**

Extreme weather events in 2021, including a

DEEP WINTER FREEZE, FLOODS, SEVERE THUNDERSTORMS, HEATWAVES, and a MAJOR HURRICANE,

resulted in annual insured losses from natural catastrophes estimated at $105 billion (USD), the fourth highest since 1970.

—Swiss Re Group

ACTION 53

PREPARE A GO BAG AND A STAY BIN

Over the last 10 years, extreme weather events have affected 1.7 billion people around the world. As climate change intensifies and changes the frequency of extreme weather events, like floods, droughts, wildfires, and hurricanes, many experts recommend putting together an emergency kit to help if you experience loss of services or damage from an extreme weather event or if you need to evacuate your home.

If you are hunkered down at home following a disaster, experts suggest having at least a two-week stash of essentials in a "stay bin." These include water, food, first-aid supplies, a hand-crank radio, a headlamp or flashlight, a solar phone battery charger, and a water filter. Other items that might prove useful are things like duct tape, a multitool, bleach wipes or paper towels, N95 masks, a whistle, and sleeping bags or blankets.

Also consider having a backup source of power. An increasing number of battery-backup power supplies can provide over 10 hours of charging for small appliances, phones, laptops, or fans. These battery backups are an alternative to gasoline-powered generators, which must be used outdoors and require regular maintenance. Consider adding a solar panel to a battery-backup power supply for a more reliable way to charge phones or to run essential health equipment like a CPAP machine.

In addition to creating a stay bin, experts advise putting together a "go bag" that you can quickly grab in the event of an evacuation. This kit should be packed with critical documents like IDs or passports, birth certificates, cash, phone chargers, energy bars or other snacks, water, medications, spare glasses, vaccination records for pets (in case you need them for accessing a kennel, park, or hotel while away from home), and a small supply of pet food. Depending on how you organize your gear, you could also add a spare set of clothes for each person in your family and some toys for any children. If you have small children, add diapers, wipes, and formula as needed.

▸ **Are you ready to weather the storm? What's in your emergency kit?**

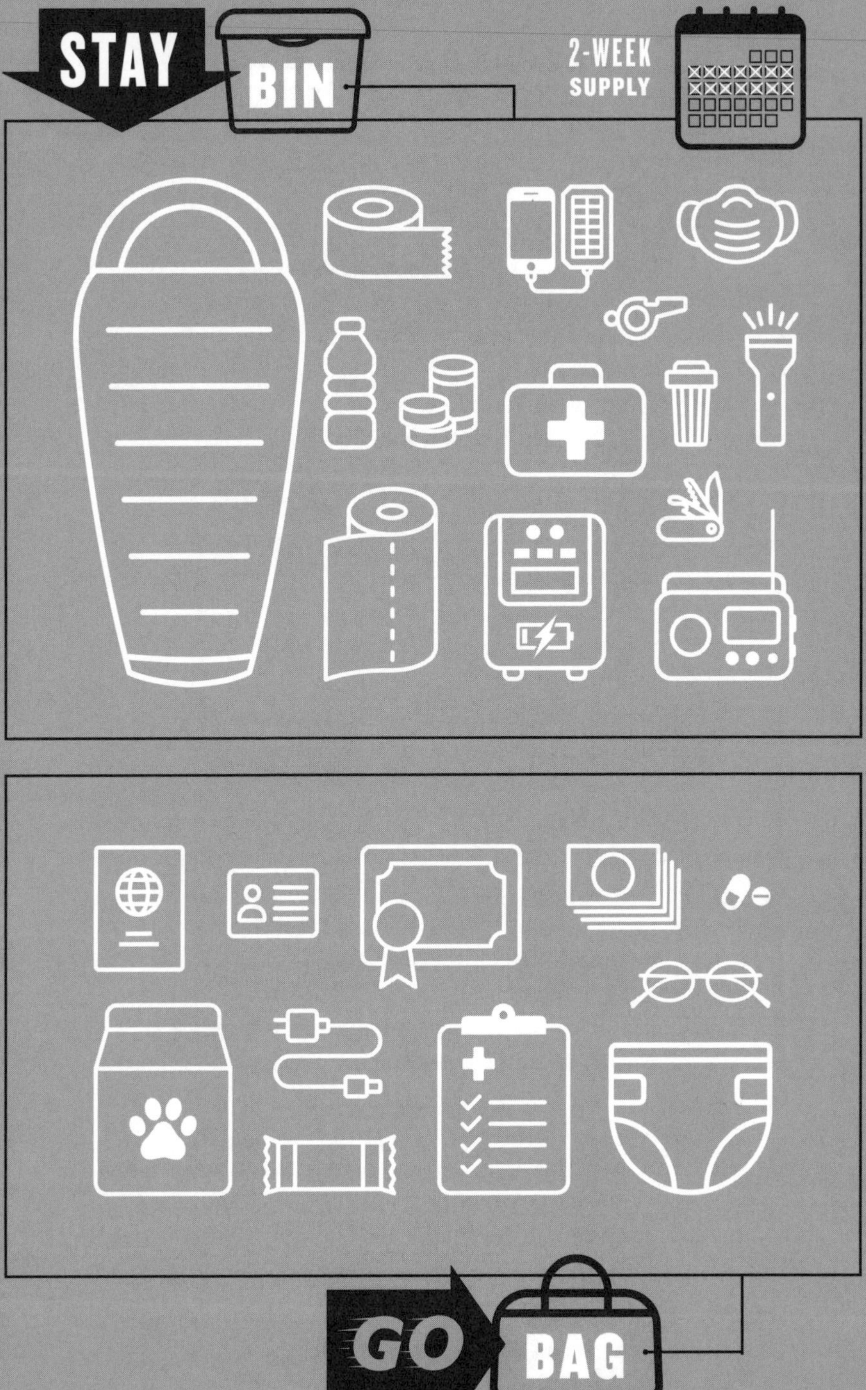
STAY
BIN
2-WEEK
SUPPLY
GO
BAG

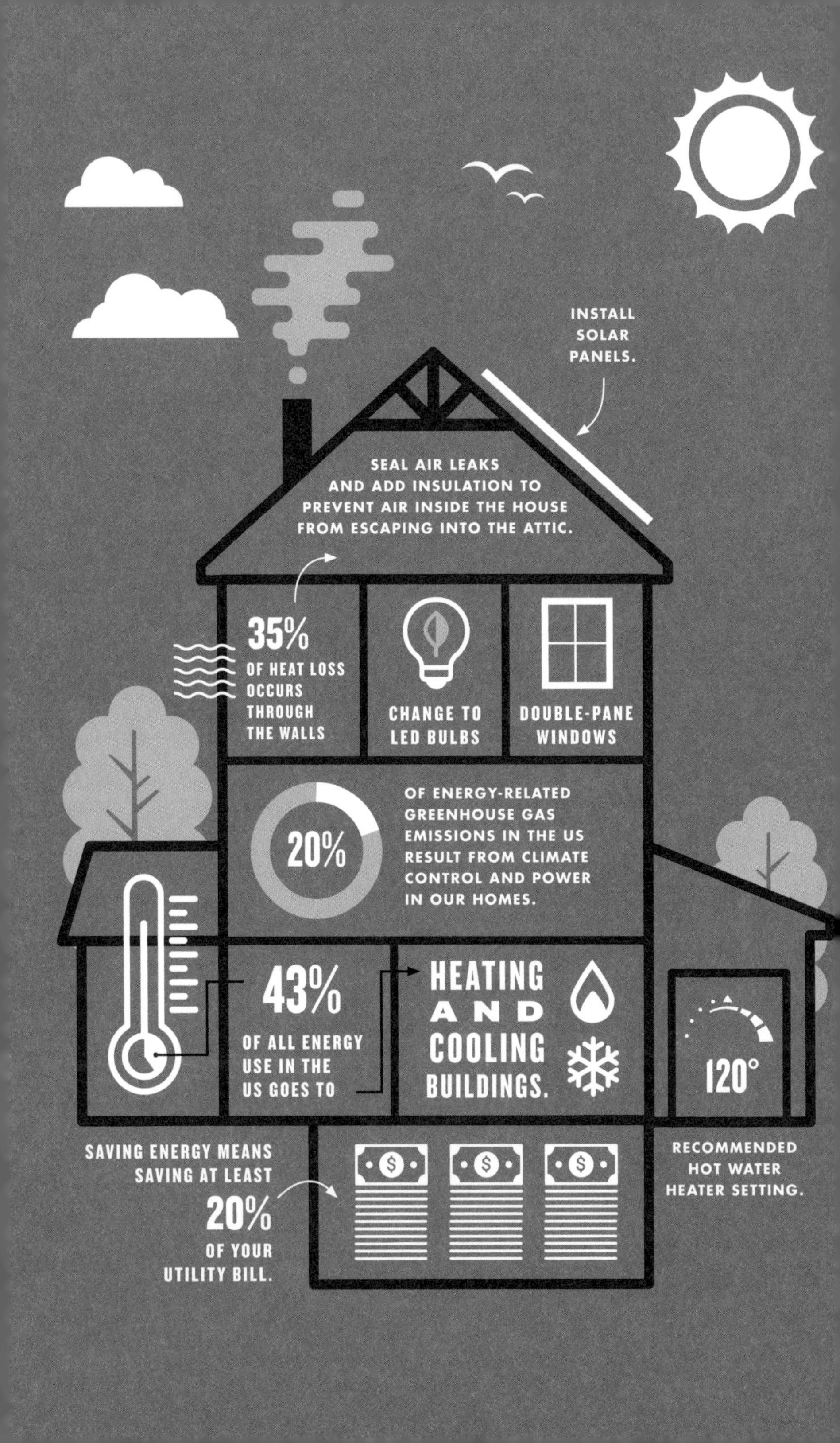
INSTALL SOLAR PANELS.
SEAL AIR LEAKS AND ADD INSULATION TO PREVENT AIR INSIDE THE HOUSE FROM ESCAPING INTO THE ATTIC.
35% OF HEAT LOSS OCCURS THROUGH THE WALLS
CHANGE TO LED BULBS
DOUBLE-PANE WINDOWS
20%
OF ENERGY-RELATED GREENHOUSE GAS EMISSIONS IN THE US RESULT FROM CLIMATE CONTROL AND POWER IN OUR HOMES.
43% OF ALL ENERGY USE IN THE US GOES TO
HEATING AND COOLING BUILDINGS.
120°
RECOMMENDED HOT WATER HEATER SETTING.
SAVING ENERGY MEANS SAVING AT LEAST 20% OF YOUR UTILITY BILL.
$
$
$

ACTION 54

CREATE A MORE ENERGY-EFFICIENT HOME

Whether you live in a single-family home, multifamily housing, a manufactured home, or assisted living, rent, own, or are a landlord, we can all work to improve the energy efficiency and overall comfort of the places we call home. Did you know that 43 percent of all energy use in the United States goes toward heating and cooling buildings? Further, energy used in the United States for climate control and power to our homes accounts for approximately 20 percent of the nation's energy-related greenhouse gas emissions! For context, that's equivalent to the total emissions produced by the entire country of Brazil. Clearly, energy efficiency investments in our homes can have a real climate impact.

Improving the energy efficiency of your home isn't only an environmentally friendly action; it can actually save you money! Energy-efficient upgrades like adding insulation, limiting the drafts from windows and doors, using smart thermostats, and maximizing sunshine can all lead to a reduction in overall energy use, fewer emissions of greenhouse gases, and a lower energy bill. The climate and cash savings can seriously add up.

Make sure you regularly maintain and clean your heating and cooling systems. When you are looking to replace boilers, heaters, or air-conditioning units, take a look at energy-efficient models and consider transitioning away from systems that rely on natural gas, propane, or fuel oil and toward efficient electric models. High-efficiency heat pumps, which are heating and cooling systems powered by electricity, are increasingly popular in the United States and are widely used throughout Europe and around the world. To sweeten the deal, many energy companies and state and federal governments offer incentives that reduce the overall price of installing energy-efficient equipment or reduce the costs associated with weatherproofing and increasing the energy efficiency of your home.

▸ **What energy improvements will you make?**

ACTION 55

ELECTRIFY YOUR HOME, FROM COOKTOP TO ROOFTOP

Around the world, 40 percent of the global population uses fuel sources like wood, charcoal, dung, and coal as primary cooking fuels, which create harmful pollutants associated with 4.3 million premature deaths each year. In the United States, around 66 percent of households use electricity to power stovetops, with around 35 percent using natural gas. With 92.4 million American households cooking one to several hot meals a day, kitchen appliances get some real use but can also be a cause for concern.

Did you know that the air we breathe indoors can be, on average, two to five times more toxic than the air outside, in part due to the off-gassing of products in our homes and from gas stoves and appliances? As we work toward more airtight and well-insulated homes (see Action 54), one consequence can be poorer indoor air quality, especially if our homes run on gas furnaces, gas hot water heaters, or gas stoves. Gas stoves can actually release some of the same pollutants as our cars, making cooking at home potentially hazardous to our health. In fact, one study indicated that homes with gas stoves can have concentrations of pollutants like nitrogen dioxide 50 to 400 percent higher than homes that use electric stoves. This can directly increase the risk of asthma and other respiratory diseases and create air quality that would be considered illegal even outside. In fact, kids in homes with gas stoves have a 24 to 42 percent higher risk of having asthma. Clearly, a push for cleaner stoves and fuels and electrification of appliances the world over could translate to improved health and a reduction in emissions of pollutants.

A switch to electric, however, can come with trade-offs, from higher up-front expense to total cost of operation, depending on the price of electricity in your region. Increasingly popular are induction cooktops and ranges, which use an electromagnetic field to transfer energy directly from the cooking surface to the cookware causing it to heat up. These induction cooktops can be much more energy efficient and even faster, shaving off two to four minutes when boiling water!

Beyond stoves, there are options available to electrify our hot water systems and provide more efficient heat, including high-efficiency heat pumps. While still requiring a significant investment, research suggests that the savings and climate benefits can add up. As someone who lives in an older home, I know that some of these transitions can be difficult. Opting to go fully electric in my 1905 house requires a new electrical panel and expanded electrical capacity in our home, a major financial investment. We saved for years before being able to upgrade to high-efficiency heat pumps, and we are still saving to make the switch from our other fossil-fuel gas appliances.

Without more incentives, tax breaks, and reductions in cost, these upgrades are out of reach for many people. This is even more true for low-income populations who tend to have greater exposure to poor indoor and outdoor air quality. It is, therefore, imperative that we push elected officials and those establishing building codes to ensure these "upgrades" become standard features of all homes, new buildings, and commercial construction. Healthy air at home (and outside) shouldn't be a privilege but a right.

▸ **How can you switch to electric in your home? Will you help advocate for clean air for all members of your community?**

ACTION 56

BE THOUGHTFUL ABOUT AIR-CONDITIONING

In 2019, around 1 billion metric tons of carbon dioxide emissions were produced cooling our homes, workplaces, stores, and industrial facilities around the world. That's equivalent to the greenhouse gas emissions produced by over 215 million gasoline-powered passenger vehicles driven for a year. Globally, between 1990 to 2016, the energy associated with cooling buildings tripled. The market for air-conditioning in emerging and developing economies is also heating up, with an estimated 10 percent increase in demand for cooling equipment between 2018 and 2019 alone!

As the world heats up, the global demand for air-conditioning is expected to triple by 2050, with an estimated 66 percent increase in the number of cooling units installed by 2030. Ensuring a sufficient, reliable, equitable, and clean supply of energy and climate-controlled spaces is critical. This will require a comprehensive response by governments, corporations, and local and state decision-makers. According to the International Energy Agency, increasing the energy efficiency of our air-conditioning systems could avoid up to 460 billion tons of emissions over the next 40 years and save around $3 trillion over the next 30 years.

As an individual, you can take some important steps to reduce cooling-related emissions: Replace systems older than 10 years with new, high-efficiency models, install a smart thermostat, avoid making your house an icebox, and weatherproof your home to keep cool air inside. Consider engaging with your local elected officials to ensure that everyone in your community has access to cool public spaces, like cool and clean air shelters during heat events. We also need as many voices as possible advocating for energy efficiency and energy security at local, federal, and international levels.

So be a cool kid on the block and turn on your high-efficiency AC unit sparingly. Then write an email or letter to an elected official asking for their commitment to energy efficiency and energy security.

▸ **How will you choose to cool down?**

BY 2050, NEARLY 2/3 OF THE WORLD'S HOUSEHOLDS COULD HAVE AN AIR CONDITIONER

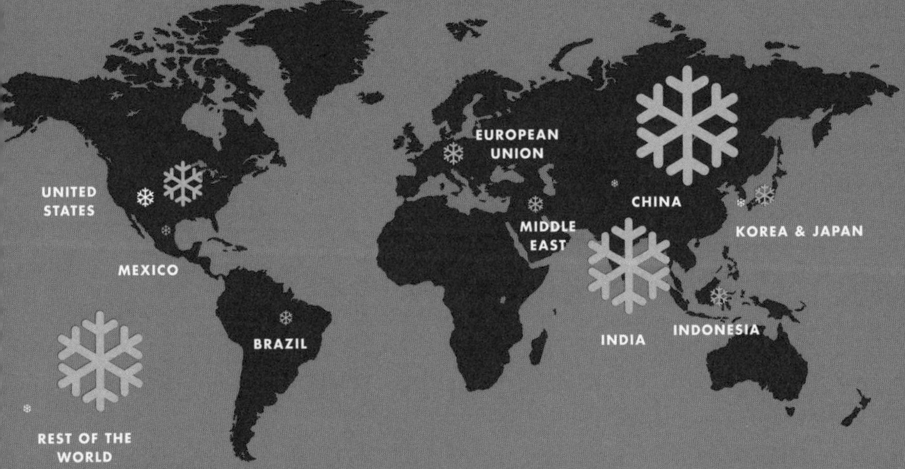

NUMBER OF AIR CONDITIONERS IN 1990

NUMBER OF AIR CONDITIONERS PROJECTED FOR 2050

SHARE OF POPULATION LIVING IN A HOT CLIMATE

SHARE OF POPULATION THAT OWNS AN AIR CONDITIONER

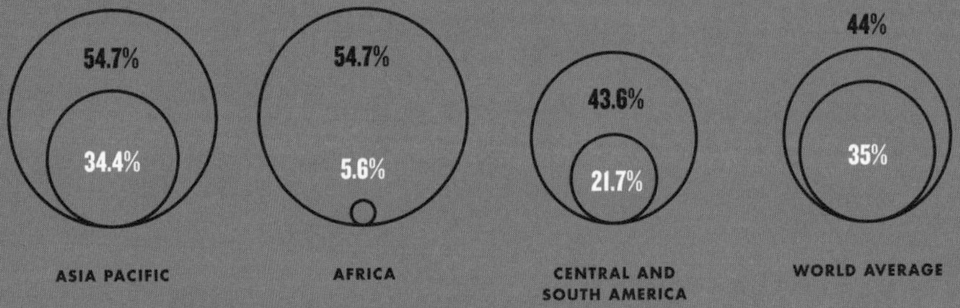

TEN COUNTRIES ACCOUNT FOR ALMOST 80% OF ALL RENEWABLE CAPACITY GROWTH.
CHINA 43%
USA 12%
INDIA 7%
GERMANY 3%
JAPAN 2%
BRAZIL 2%
SPAIN 2%
FRANCE 2%
KOREA 2%
AUSTRALIA 2%
OTHER 23%

ACTION 57

GO SOLAR

Despite increasing transport and commodity prices, the International Energy Agency (IEA) expects record growth in renewable energy. This is driven, in part, by increasingly strong government policies and clean-energy goals set at the end of 2021. Global renewable electricity is slated to grow more than 60 percent by 2026! Wind energy capacity is set to increase by 25 percent by 2026. It might be surprising to learn that even with this faster deployment, experts indicate that even greater adoption of renewable energy is needed to achieve global net-zero emissions by 2050.

In 2020, solar photovoltaic (PV) technology provided the "cheapest electricity . . . in history," according to the IEA. Solar PV currently accounts for 60 percent of new renewable energy capacity being added to the electric grid. Most of this is being added in large-scale systems by electric utilities. So what about solar panels on your home, apartment building, business, or workplace? These installations, which differ from utility-scale systems, are often referred to as distributed solar PV. Such systems are also an important part of the renewable energy story. Distributed systems are becoming increasingly cost-effective. In the United States, a range of incentives are available—from federal tax incentives to utility-provided rebates—to reduce the up-front cost of installation. In 2015, distributed systems accounted for 30 percent of solar PV in the world. However, integration with the grid continues to be a barrier to these installations in some regions.

Globally, the expansion of distributed solar PV could pack a real climate and cost-savings punch. As outlined in Action 8, Project Drawdown suggests that increased uptake of distributed solar PV over the next 30 years could avoid anywhere between nearly 28 to 69 gigatons of greenhouse gas emissions and save up to more than $13 trillion in operational costs over the lifetime of the systems.

If you don't have a rooftop suitable for solar, consider installing a solar hot water system or high-efficiency hot water heating system in lieu of one run by natural gas. Did you know that solar hot water can reduce energy use by anywhere from 50 to 70 percent?

▸ **Will you consider going all in on solar in any way you can?**

ACTION 58

LIGHT WITH LEDS

LEDs, or light-emitting diodes, are increasingly common; they are in our home lightbulbs, strings of holiday lights, stoplights, and more. Aside from the environmental benefits of installing LED lighting, it can lead to more green in your wallet. According to the US Department of Energy, LED lights use around 90 percent less energy and have a life span around 25 times longer than a comparable incandescent lightbulb. With lighting in our homes accounting for about 15 percent of our typical home electricity consumption, a switch to LED lightbulbs can lead to an average savings of $225 per year! On a per-bulb basis, an LED bulb can save you anywhere from $50 to $100 over its lifetime. Fortunately, across the United States there is wide adoption of LED lighting, and by 2035, LEDs are expected to be used across a majority of commercial lighting in the country.

According to the EPA's household carbon footprint calculator, if my household replaced just five incandescent bulbs with an Energy Star–approved lightbulb, like an LED, we'd save $20 annually on our energy bill and 255 pounds of CO_2 emissions, about the equivalent of that produced by 13 gallons of gas or burning 128 pounds of coal! (Don't worry: our whole home is already on LEDs, both for the climate and the cost savings!) When we zoom out to the larger scale, the broader global adoption of LEDs over the next 30 years would result in some eye-popping numbers: in residences alone, LEDs could lead to an estimated 10.2 to 10.8 fewer gigatons of CO_2 emissions! It would take 2,772,533 wind turbines running for an entire year to avoid 10.2 gigatons of CO_2 emissions!

▸ **Will you lighten the load by lighting with LEDs?**

Lighting accounts for around 15% of an average home's electricity use, and the average household saves about $225 in energy costs per year by using LED lighting.

—US Department of Energy

ACTION 59

GO LOW-FLOW WITH YOUR FIXTURES

The leading climate solutions research organization, Project Drawdown, put low-flow fixtures, like taps and showerheads, which help reduce wasted water, on its list of high-impact household actions. Water use could increase 270 percent globally by 2050, so the installation of low-flow fixtures could bring multiple benefits—emissions reductions, cost savings, and protection of vital water resources. As water supply issues are likely to worsen as the world warms, going low-flow with our fixtures is critical.

Putting water stress in context, the Intergovernmental Panel on Climate Change estimates that with 1.5°C (2.7°F) of warming relative to the preindustrial period, more than 4 percent of the global population will experience new or worsened water scarcity, and over 114 million people will be exposed to drought each month. With the current rate of warming, we could pass 1.5°C in the early 2030s.

What kind of savings are we looking at from low-flow taps and showerheads? Wide adoption globally (81 to 92 percent) over the next 30 years could translate into avoiding 1 to 1.6 gigatons of greenhouse gas emissions (CO_2e) and $468 billion to $765 billion in water heating saving costs. While there is an up-front cost to implementation, today is the day to go low-flow with your fixtures.

Before you go out and buy new fixtures, ask your local municipality or utility provider if they have any discounts or free kits. After researching this in my community, I was able to get a set of free low-flow aerators for my sinks and two free low-flow showerheads from my power company!

As with most things in this book, action across scales and sectors is also needed. For example, water conservation, efficiency, and reduction efforts are needed across high-water-use sectors like agriculture. The agricultural sector is the primary water user in 50 percent of US counties. Efforts to reduce water use are essential in all sectors whether agriculture, industry, at work, and at home.

▸ **Will you opt to go low-flow? Will you advocate for water conservation in your community?**

WATER USE

COULD INCREASE

270%

GLOBALLY BY 2050,

so the installation of low-flow fixtures could bring multiple benefits—emissions reductions, costs savings, and protection of vital water resources.

WATER HEATING REPRESENTS ABOUT 18%
OF OUR TOTAL ANNUAL HOUSEHOLD ENERGY CONSUMPTION

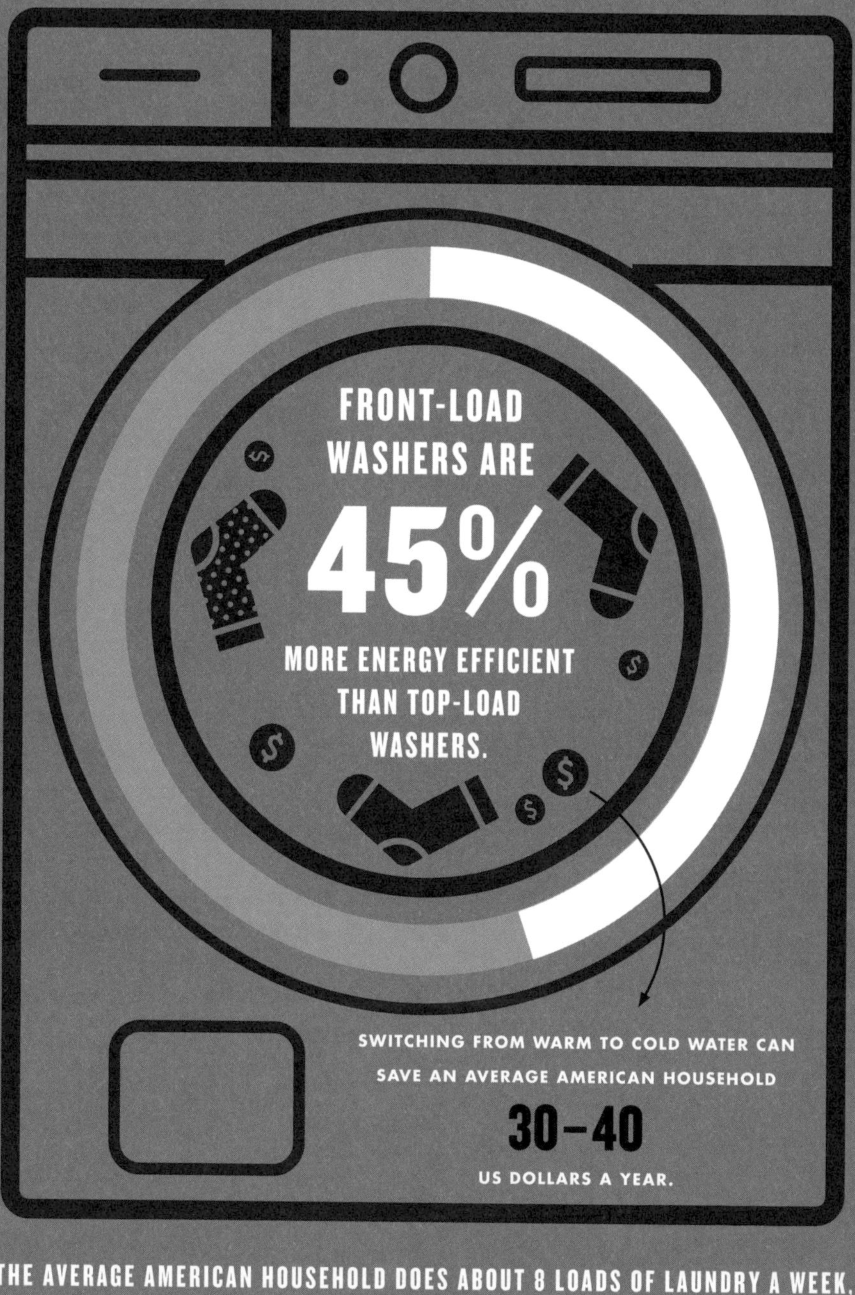

FRONT-LOAD WASHERS ARE
45%
MORE ENERGY EFFICIENT THAN TOP-LOAD WASHERS.
SWITCHING FROM WARM TO COLD WATER CAN SAVE AN AVERAGE AMERICAN HOUSEHOLD
30–40
US DOLLARS A YEAR.

THE AVERAGE AMERICAN HOUSEHOLD DOES ABOUT 8 LOADS OF LAUNDRY A WEEK,
TOTALING AN AVERAGE OF 416 LOADS EACH YEAR!

ACTION 60

CLEAN YOUR CLOTHES EFFICIENTLY

The average American household does about eight loads of laundry a week, totaling an average of 416 loads each year! Heating water—which includes heating water for our laundry—represents about 18 percent of total annual household energy consumption.

In 2015 and 2018, the United States updated its federal standards for water and energy efficiency for household, multi-unit household, and laundromat washing machines. These standards have led to the average washing machine using 75 percent less energy than machines produced in 1987—all despite a near doubling in the size of the average machine over that time. These standards are expected to result in an estimated $30 billion in savings for consumers, save 3 trillion gallons of water, and save the equivalent energy to that produced by two average coal plants over 30 years.

When it is time to replace the washing machine in your household, consider a front-loading machine that is Environmental Protection Agency (EPA) Energy Star–certified. According to the EPA, "Energy Star–certified front-load washers are 50 percent more water efficient and about 45 percent more energy efficient than top-load washers with agitators." As an added bonus, front-loading machines are also better at cleaning your clothes!

You can further reduce your impact by washing your clothes on shorter cycles, only washing full loads, using cold water, and opting to hang dry your clothes. Switching from hot to warm water can reduce energy use by half while a switch to cold can yield even greater energy savings. That switch to cold can also save some cold hard cash—with an average US household saving $30 to $40 every year just by washing with cooler water.

▶ **The next time you do your laundry, what temperature and wash cycle will you choose?**

GARDEN FOR A GREENER PLANET

As we consider climate change in our homes, we should also consider our yards, gardens, and green spaces. Many gardeners are likely familiar with the USDA's plant hardiness zones, which are based on the average annual minimum winter temperature, and are often used to determine which plants are most likely to thrive in a given region. In a warming world, plant hardiness zones have shifted northward, changing the types of trees, flowers, shrubs, and other plants that thrive in different regions.

As warming continues, selecting plants adapted to new climate conditions will help ensure a thriving and resilient garden. This is particularly true for trees, which are increasingly showing a northward shift in their range. One study documented that across 30 eastern US states, 70 percent of the tree species assessed were migrating northward. Want to know what might thrive in your area? Reach out to your local extension service for information and advice. Other climate-friendly gardening options include planting perennials and pollinator-friendly plants and using less turf grass.

Interested in documenting the changes you see in your yard or community? The USA National Phenology Network is always looking for observers around the country to help track the timing of seasonal changes in plants—like when cherry blossoms or lilacs bloom or when the fall colors arrive. These locally collected data are critical for helping to track changes across plant and animal communities.

When it comes to the lawn and gardening tools we use, opting for electric is key. Take gas leaf blowers as an example. Those seemingly small devices pack a real climate punch. A consumer-grade leaf blower emits more pollutants than a large truck! In fact, these tiny but mighty gas guzzlers may soon be outlawed; California has plans to outlaw such machines as early as 2024.

In addition to making a shift in the plants you grow at home, consider encouraging your local parks department and other yard service providers to make the shift as well. It can be good for their wallets, their workers, and all of our health, not to mention our climate!

▸ **Will you adapt your yard and garden and go green with your equipment?**

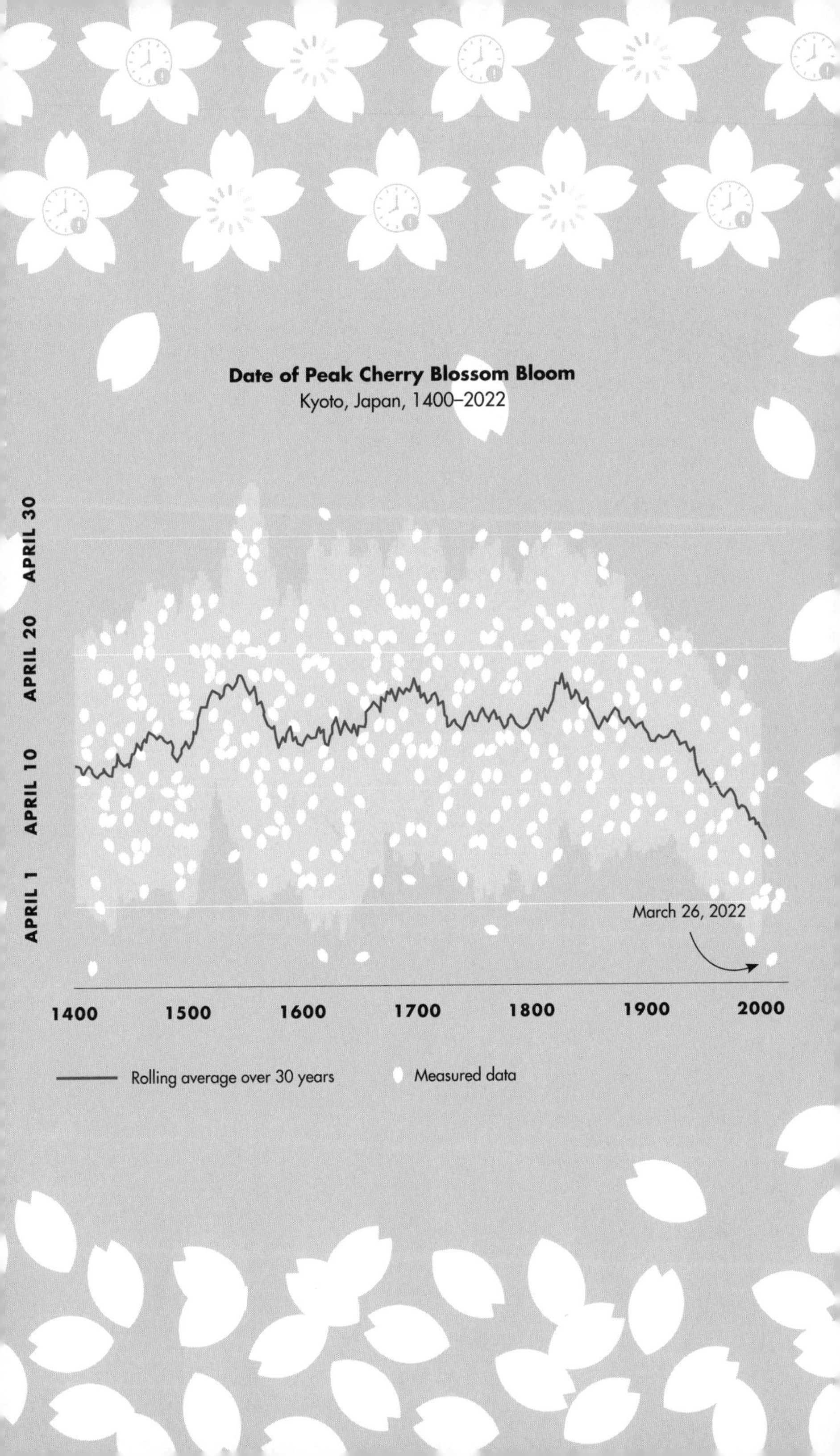
Date of Peak Cherry Blossom Bloom
Kyoto, Japan, 1400–2022
APRIL 30
APRIL 20
APRIL 10
APRIL 1
March 26, 2022
1400
1500
1600
1700
1800
1900
2000
Rolling average over 30 years
Measured data

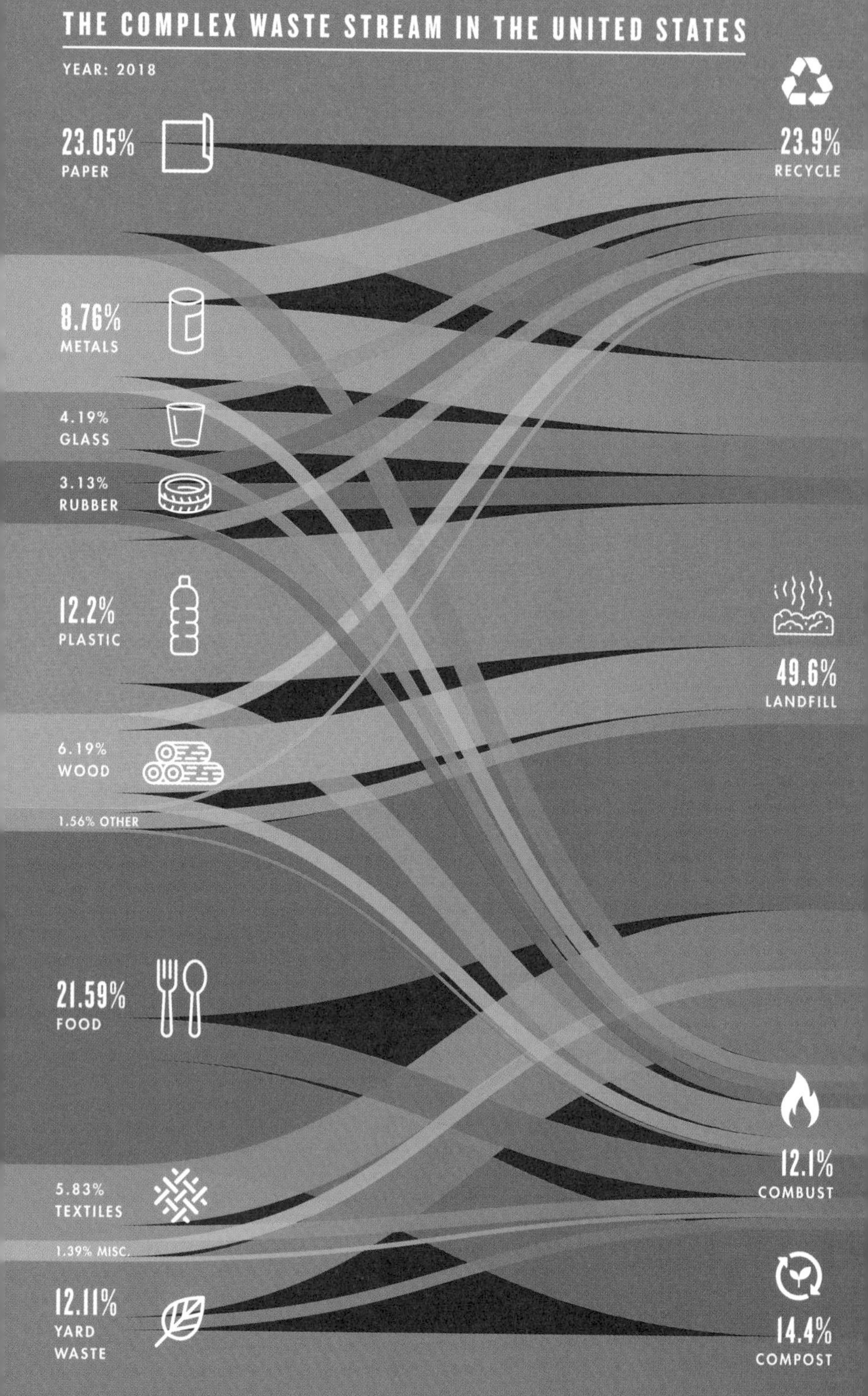
THE COMPLEX WASTE STREAM IN THE UNITED STATES
YEAR: 2018
23.05% PAPER
8.76% METALS
4.19% GLASS
3.13% RUBBER
12.2% PLASTIC
6.19% WOOD
1.56% OTHER
21.59% FOOD
5.83% TEXTILES
1.39% MISC.
12.11% YARD WASTE
23.9% RECYCLE
49.6% LANDFILL
12.1% COMBUST
14.4% COMPOST

ACTION 62

REDUCE WASTE AND RECYCLE

We love our waste. Well, maybe we don't love it, but we sure do create tons of it. In fact, the United States produces hundreds of millions of tons of waste each year. In 2018, it produced nearly 36 million tons of plastic waste with about five pounds of total waste produced by each person every day. More than 75 percent of that plastic waste went to the landfill. As of 2019, an estimated 40 percent of households don't have access to recycling of any kind. So it may be unsurprising that less than 10 percent of plastic produced in the United States is recycled.

Why? Recycling is expensive, virgin materials can be cheaper than recycled ones, and China, a major market for US recyclables, is no longer accepting this waste. Recent investigative journalism also uncovered that the plastic industry intentionally misled the public into thinking that plastic recycling worked well. Why? To sell more plastic goods.

Of course, other materials like paper, glass, and metal can be recycled. Cardboard and paper accounted for nearly 67 percent of all recycled materials in the United States in 2018. If paper recycling adoption increased even slightly from current-day levels, there could be up to a 7 percent reduction in annual emissions from the paper industry. Even better, aluminum recycling can use up to 95 percent less energy than creating new aluminum products.

So what can you do? Reduce the amount of waste you generate. Be thoughtful about your recycling. Avoid "wish cycling," by tossing something you hope is recyclable but in reality might muck up the system or contaminate otherwise recyclable waste. Learn more about your local recycling process: Do you really know what happens to the trash and recycling you generate? We also need to advocate for more comprehensive and economically viable recycling options. About 50 percent of waste globally is industrial, so expansion of recycling efforts for all waste streams is essential. New waste fee structures, redeemable deposits, advances in processing, reduction in packaging use, and improved industry regulation are all likely to help accelerate both residential and commercial recycling.

▸ **How will you clean up your garbage and recycling efforts?**

ACTION 63

CALCULATE YOUR CARBON FOOTPRINT

The shifting of responsibility and blame to individuals to solve the climate crisis is a narrative many of us are familiar with, and one that, in part, led to much of the research highlighted in this book. As we know, avoiding the most negative consequences of a changing climate requires large, systems-based changes, many of which are out of the direct control of individuals—say, revising building codes and design standards, adopting utility-scale renewable energy, and reducing the supply chain emissions of a large multinational corporation. Therefore, throughout this book, individual actions are supplemented by clear calls for collective action.

That said, many people still want to understand their personal or individual contribution to greenhouse gas emissions. Carbon footprint calculators are one tool to help you understand your household's major sources of greenhouse gas emissions, as well as where you might have some easy wins in reducing emissions. Tools like the Environmental Protection Agency's household carbon footprint calculator can be a helpful first step in your climate action journey, and a touchpoint throughout.

Don't forget that in the pursuit of a cleaner, greener, and more equitable world, there is more than carbon emissions to truly quantify our impact and for identifying opportunities for action. Supporting community efforts to build resilience to climate change, strengthening social ties in your community, building a close community of colleagues at work, and flexing your creative muscles to motivate, sustain, and expand your broader climate work should all factor into your personal equation for your "footprint".

The world needs each and every one of us to bring our strengths, creativity, and ideas to this work. So while doing what you can to reduce your carbon footprint certainly counts and is critical, so is a well-balanced portfolio of climate work. This includes advocating for and supporting the systems-based change we know is required to move us all toward a better, brighter future.

▸ **What's your carbon footprint? Now, what's your climate action footprint?**

A

"CARBON FOOTPRINT"

is one tool for understanding our impact, but a full and balanced portfolio of climate work is needed—from emissions reduction activities to efforts that enhance individual and community climate resilience.

WHAT'S YOUR CLIMATE ACTION FOOTPRINT?

NATURE-BASED AND NATURAL SOLUTIONS

ACTION 64

UNDERSTAND THE PROSPECTS OF CARBON REMOVAL

Let's be clear: the surest way to avoid the worst consequences of a changing climate is to reduce emissions of greenhouse gases across all sectors, as quickly as possible. Scientists have given us a budget for these emissions, but with each passing day our carbon budget is increasingly heading toward the red. Along with the actions outlined elsewhere in this book, another tool for balancing this budget is land- and ocean-based carbon dioxide removal (CDR) techniques.

CDR can span everything from planting trees, keeping forests and coastal wetlands intact, storing carbon in soils, and employing technological innovations that remove carbon from the atmosphere and store it. However, like with any good budget, we have to understand our spending habits, the costs and benefits of different actions, and, ultimately, where to find the biggest bang for our buck.

While we shouldn't hold out hope for high-tech fixes and yet-to-be-proven techniques to "save us," the Intergovernmental Panel on Climate Change (IPCC) notes that we need to reduce emissions while also scaling up our preservation and restoration of natural landscapes like forests and wetlands and working toward the deployment of engineered systems that can capture and store emissions produced by our activities. In fact, the IPCC says all pathways to limit warming at or below our global goals require "rapid and profound near-term decarbonizations of energy supply" and some amount of removal of atmospheric carbon dioxide. In short, if we want to avoid surprises and potentially hard-to-deploy solutions that will come with their own social, economic, and ecological impacts, we are better off working to avoid emissions at their source, while using solutions and carbon dioxide removal as complements to emissions reductions.

In many cases, technological approaches to CDR are not yet proven at scale. As of 2021, only 19 systems exist that directly capture CO_2 from the air. Many countries, companies, and philanthropists are making large investments to help accelerate future development of these technologies. The CDR techniques that are now "proven" at scale are primarily natural solutions, like reforestation or afforestation (putting forests where they weren't previously).

As an individual, you may be offered options to purchase "carbon offsets" to reduce the impact of certain activities like flying. These programs often claim to deploy CDR techniques, like planting trees, to remove—or offset—the emissions you (or a company, state, or country) create. However, many of these offset programs are not regulated, so it can be hard to track the quality or integrity of such commitments. When you are considering buying offsets for your activities, research the company providing the offset (including what CDR activity is actually being promised), verify that the project actually exists and that you aren't funding a yet-to-be-started project, and check that the program has a form of third-party certification (see Action 67).

▸ **What emissions will you avoid producing?**

ACTION 65

LEARN ABOUT BIOENERGY AND CARBON CAPTURE AND STORAGE

In the jargon-laden language of climate change, you may have heard of bioenergy, as well as carbon capture and storage. These two terms can often be found combined together to refer to bioenergy with carbon capture and storage, or BECCS. Let's unpack this a bit more. (And to be clear, this topic itself could be an entire book!) Bioenergy is energy created by processing plant- and biological-based materials, like trees and woody debris, algae, crops like corn and sugar, grasses, and even food waste.

Collectively, these materials are often called biomass. Because biomass can be regrown, it is considered a renewable resource. Bioenergy is produced through a range of methods, from burning biomass directly for energy (like trees) to converting biomass to liquid fuels like ethanol, or using fermentation. Biofuels are widely used today as part of the blend of vehicle fuel, with gasoline in the United States consisting of about 10 percent ethanol by volume.

With a call to expand bioenergy, what are some primary challenges of expanding the use and development of bioenergy? It can require significant areas of land and sometimes fertilizer- and water-intensive practices. The production of this energy can also generate emissions.

BECCS, however, is one solution to the emissions problem. This is where bioenergy generation is paired with carbon capture and storage technologies, which capture any emissions generated from the process and store them deep underground. This, in theory, can make the process carbon negative because some of the carbon in the plant is never released back into the atmosphere, meaning that more emissions were removed than created by the production of bioenergy.

The Intergovernmental Panel on Climate Change assumes that expanded use of BECCS will be included in global efforts to meet emissions-reduction targets to limit global warming, but they also note that bioenergy comes with benefits as well as potentially "adverse side effects," like land degradation, food insecurity, stresses on ecosystems, and biodiversity loss.

The development of advanced biofuels, like the use of algae, may help to reduce some of these impacts, but biofuels are generally considered a bridge to help us span the gap between current carbon- and emissions-intensive energy and a fully decarbonized energy supply. This is particularly true for harder-to-decarbonize sectors, like aviation and cargo shipping, but more research, development, and scaling of these technologies is needed.

In 2021, the International Energy Agency (IEA) noted that the role of biofuels was a key uncertainty in the pathway to net-zero emissions by 2050, but also that reaching these goals with our energy system "requires the immediate and massive deployment of all available clean and efficient energy technologies." So, on the whole, we need to keep a sharp eye on emissions-reduction opportunities, and work to limit land degradation and other trade-offs of biofuels, while not becoming overly reliant on hard-to-scale or yet-to-be-proven technologies and systems.

The IEA also notes that "a transition of such scale and speed cannot be achieved without sustained support and participation from citizens, whose lives will be affected in multiple ways." How can we as individuals engage? Part of our role in supporting the advancement of bioenergy and carbon capture and storage can include expanding its use in our communities' buildings, buses, and local energy systems and advocating for expansion of efforts to develop and scale the most efficient forms of bioenergy.

▸ **Will you help advance the use of BECCS in your community and help champion additional research in this space?**

THE EARTH HAS 4.06 BILLION REMAINING HECTARES OF FOREST.

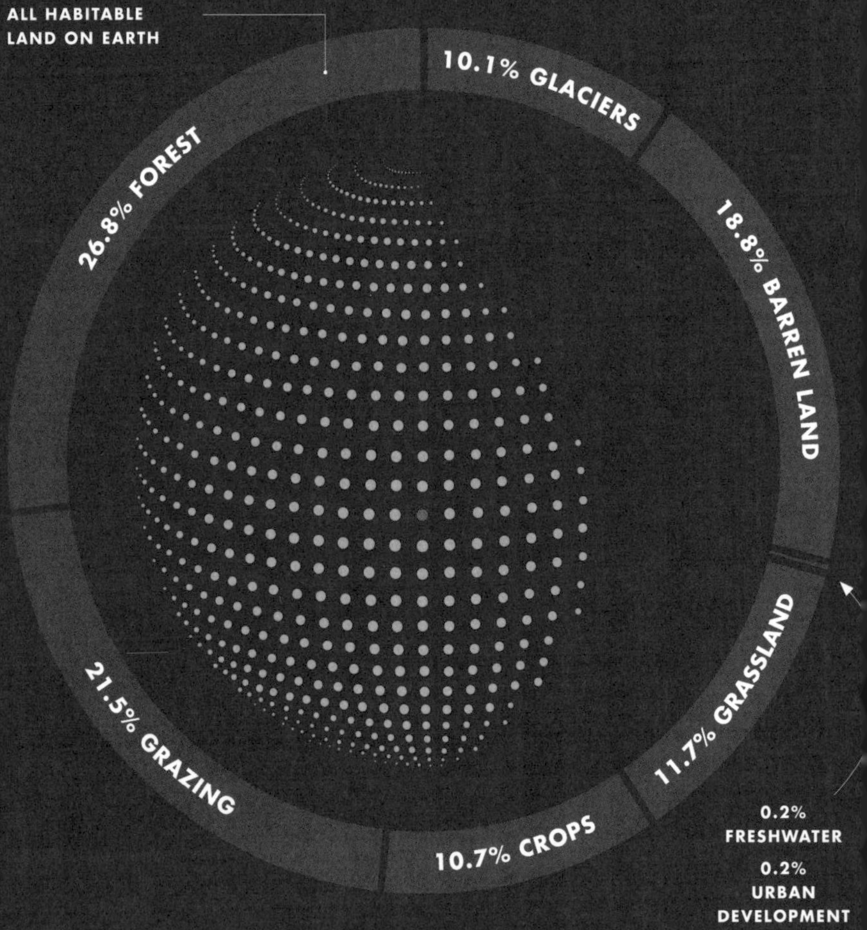

7.3

MILLION HECTARES OF FOREST: THE AMOUNT OF FOREST LOST EVERY YEAR

ONE

BILLION HECTARES IS EQUAL TO AN AREA THE SIZE OF THE UNITED STATES

ACTION 66

HELP KEEP FORESTS HEALTHY AND INTACT

Forests are considered "carbon sinks," meaning they store or absorb more carbon than they produce. Trees consume carbon dioxide through the process of photosynthesis—the carbon from CO_2 in the air is stored in the tree as wood. This natural process acts as a sink, with the carbon staying in the tree until it dies and decomposes or is burned. In the United States, forests have stored about 14 percent of the country's CO_2 emissions!

While wildfires, deforestation, or large stands of dead trees can turn forests into carbon sources instead of sinks, restoring forests, keeping our existing healthy forests intact, and planting more trees can help us to increase the capacity of this natural carbon sink. In fact, trees are a critical tool in meeting our global climate goals, with some estimates indicating that Earth could support about 25 percent more forested area than now covers the landscape. This would require planting more than 500 billion trees, which would serve as a sink for about 25 percent of current atmospheric carbon. This would be equivalent to negating about 20 years of human-generated carbon emissions.

So shouldn't we just plant trees and call it a day? Well, where we plant trees, what trees we plant, and other benefits of trees for people and biodiversity are all worth unpacking further. Read on (see Actions 67, 70, 72), and then go put down some (tree) roots!

▸ **Will you support the preservation and health of the world's forests?**

ACTION 67

CAREFULLY CONSIDER CARBON OFFSETS

Swipe your credit card . . . your credit card company will plant a tree. Take a flight . . . buy an offset through planting trees! We may think a tree or two can make up for our own high-emitting activities like flying overseas, but there is more to the story. While there is no question that we need to seek natural solutions like reforestation to help achieve our global emissions reduction targets, we also can't let offsets or carbon credits lull us into a sense of major accomplishment if these offsets are not matched with real reductions in greenhouse gas emissions.

Several organizations have also studied the equity implications of the amount of land needed to achieve the large-scale carbon removal required to meet the net-zero commitments made by nations around the world. What's clear is that we need to see the forest through the trees. Oxfam has calculated that the total amount of land required for planned carbon removal could potentially require space about five times the size of India, or the equivalent of all the farmland on the planet. It is expected that the demand for this amount of land will disproportionately fall on low- and middle-income countries. This, in turn, potentially drives displacement and hunger, and shifts the burden of mitigation to those who have done the least to contribute to the problem.

As individuals, we can work to reduce our own emissions and think critically about the need for that air travel, purchasing more "stuff," or wasting food. You may be offered options to purchase "carbon offsets" to reduce the impact of certain activities like flying. These programs often claim to deploy carbon dioxide removal (CDR) techniques, like planting trees, to remove or offset the emissions you produce. However, many of these offset programs are not regulated, so it can be hard to track the quality or integrity of such commitments. As in Action 64, when considering buying offsets, research the company providing the offset including what CDR activity is actually being promised, verify that the project actually exists, and check that the program has third-party certification.

▸ **How can you both reduce and offset your emissions?**

WE NEED TO MANAGE LAND IN WAYS THAT TACKLE CLIMATE CHANGE AND HUNGER TOGETHER AND STRENGTHEN THE RIGHTS AND RESILIENCE OF COMMUNITIES.

—Oxfam International, 2021

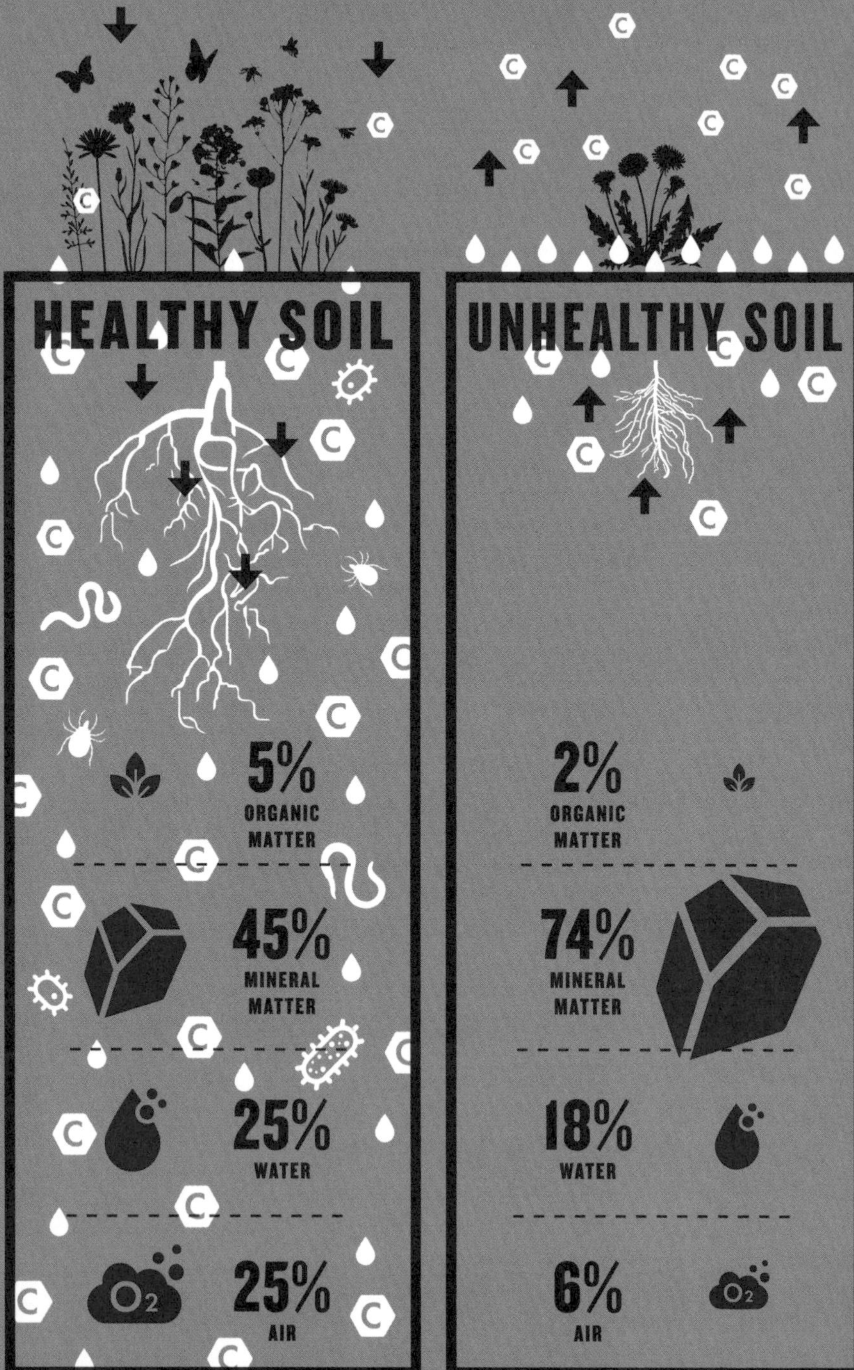

HEALTHY SOIL
5%
ORGANIC MATTER
45%
MINERAL MATTER
25%
WATER
25%
AIR
UNHEALTHY SOIL
2%
ORGANIC MATTER
74%
MINERAL MATTER
18%
WATER
6%
AIR

ACTION 68

MAKE YOUR SOIL HEALTHY

Soil is not just dirt but rather a dynamic mix of minerals, air, water, microbes, fecal matter, and organic matter—all interacting to support society as we know it. Soil provides enormous benefits, but is often underappreciated for the vital role it plays in our lives—from providing us with food and healthy ecosystems to improving water quality. Soil is also a source of climate solutions; it can sequester carbon, store water, and reduce runoff. Plants and soils currently absorb the equivalent of about 20 percent of human-produced greenhouse gas emissions.

However, declining soil health, driven primarily through land use changes like deforestation and expansion of agricultural lands, has contributed about one-third of the observed increase in CO_2 in the atmosphere, with soils in the United States having lost between 50 to 70 percent of their original organic carbon content. Improving the health of our soils can help reverse this trend, turning soils back into a carbon sink that can help us maximize their critical role as a climate solution.

According to The Nature Conservancy, improving soil health in the United States—through practices like keeping cover crops on the land to reduce exposed soils, limiting compaction and disturbance of soils, and rotating crops on all major croplands used to produce soy, corn, and wheat—could yield almost $50 million in benefits for the environment, and for society! Advocating for improved soil health practices by farmers in your region can help move the dial but many of us don't have relationships directly with farmers. So what can you do?

- Support brands that help farmers to adopt healthy soil practices.
- Ensure your yard has healthy soil: reduce disturbance and compaction, put in deep-rooted plants, and cover your soil with plants or mulch.
- Talk to your local parks department to learn what they are doing to support healthy soils. Tell them about options to plant more native plants and grasses, reduce mowing following heavy rain, and improve signage and other guidance to make sure people stay on marked paths.

▸ **Do you have healthy soil under your feet?**

ACTION 69

SUPPORT COASTAL WETLAND CONSERVATION

Coastal areas offer a rich opportunity for storing carbon. The carbon stored in these wetlands—including mangrove swamps and forests, tidal marshes, and seagrass meadows—is often referred to as "blue carbon." This blue carbon removes carbon dioxide from the atmosphere at a rate estimated to be five times greater than that of forests on land! However, degraded coastal wetland areas are less able to remove carbon from the atmosphere. So it is critical to maintain, preserve, and restore the health of these coastal areas.

Coastal wetlands naturally produce methane and carbon dioxide, but on the whole, when they are healthy and in their natural state, they remove more carbon from the atmosphere than they produce. This is only true when nothing disturbs them, or disconnects them from salt water like a road, home, or dike. Simply put, the chemistry of these coastal wetlands must be in balance for these ecosystems to work their magic. When they are in balance, with sufficient interaction with salty seawater, they can actually have a strong cooling effect on the climate through their high rates of carbon storage. But when our activities tip the balance, it can turn these natural "climate coolers" into sources of heat. In fact, human disturbances of these critical ecosystems are associated with an estimated 450 million tons of CO_2 being released each year.

According to the Intergovernmental Panel on Climate Change (IPCC), restoration of these ecosystems around the world "could provide climate change mitigation through increased carbon uptake and storage of around 0.5% of current global emissions annually." These coastal areas also provide vital habitat for birds and threatened species like the Florida panther, improve water quality that in turn helps support coastal fisheries, and provide coastal communities with storm protection. One report suggests that well-managed and protected coastal areas could reduce flood damage by as much as 29 percent!

While the IPCC has called for improved research and measurement of the potential of blue carbon, the multiple benefits of these critical coastal ecosystems are clear. If you live in a coastal community, advocate for and support effective coastal ecosystems management practices that will ensure the protection of coastal areas even as sea levels rise. This should absolutely be part of your local municipality, township, or Tribe's climate conversations. If you're a noncoastal resident, you can support organizations like The Nature Conservancy that help advance blue carbon policies.

▸ **How will you support coastal wetland preservation, conservation, and restoration?**

IF COASTAL WETLANDS WERE RESTORED TO THEIR 1990 EXTENT, THEY WOULD HAVE THE POTENTIAL TO INCREASE ANNUAL CARBON SEQUESTRATION

274 MILLION TONS

PER YEAR, OFFSETTING THE BURNING OF MORE THAN

2 BILLION

BARRELS OF OIL.

—The Nature Conservancy

ACTION 70

CONSERVE, RESTORE, AND RECONNECT LAND

Over 75 percent of our land surface has been significantly altered, and 87 percent of global wetlands have been lost over the last 300 years. Add to that a reported one million species globally that are facing extinction due to land use change, climate change, and other human impacts, and the picture is grim. According to the Campaign for Nature, only 7 percent of our global oceans and 15 percent of our land masses are protected. The need to act is urgent. We must recognize the role healthy landscapes, waterways, and habitats play in creating and supporting thriving communities for humans, plants, and animals. Our efforts to protect these critical habitats and "ecosystem services" are vital.

Take the example of habitat connectivity—as the planet warms and areas of suitable habitat shift northward and upward in elevation, it's critical that we think about how to reconnect currently fragmented landscapes to assist the movement and migration of wildlife. This ensures that they will have safe natural corridors to move into more suitable habitat. Effective land use management and planning can help shift the odds for species survival.

Thoughtful, strategic conservation and restoration approaches need to consider the true diversity of needs, uses, and functions of landscapes. This includes land use decisions made in residential areas. If you have the privilege of owning a home or building on a new parcel, consider how you might best interact with the landscape. How much of the natural land can you keep untouched? Can you provide continued access to animals and others occupying or moving through your land? If you don't own property, you can support efforts to conserve, restore, and reconnect our natural landscapes and ecosystems by volunteering with local conservation organizations, advocating for conservation and restoration at local, state, and federal levels, and through climate- and conservation-friendly investments in your local parks and green spaces.

▸ **In what ways, large or small, can you help conserve natural lands and keep them connected and healthy?**

"I think there's a lot of things that could make a resilient landscape but one of the most important things is communication and respect. And that's with everyone that lives within that landscape, because before you can get to any actions on the ground you have to have the foundation of respect and honoring and listening to one another. We need to make sure that it's not just for one, but for all."

—Amelia Marchand
Executive Director, Center for World Indigenous Studies

ACTION 71

GO GREEN WITH YOUR INFRASTRUCTURE

Did you know that green spaces and green infrastructure—like rainwater gardens, roadside plantings, green roofs, and ditches designed to let rainwater soak into the soil (a.k.a. bioswales) instead of rapidly running off—bring a multitude of benefits? By increasing the number of surfaces that can soak up water and by slowing the speed of water running into stormwater systems, green infrastructure works to mimic natural processes in areas that might otherwise be covered with impermeable surfaces like concrete. While green infrastructure is often installed to assist with the management of rainfall and stormwater, it can also improve air quality, reduce erosion, limit pollutants entering our lakes, rivers, and beaches, and provide important habitat for birds and insects. It can help to reduce the impacts of climate change in our communities, from rising seas to increasing rainfall and extreme heat.

Green infrastructure is the counterpart to "gray infrastructure," the more traditional forms of infrastructure that are engineered by humans like dams, seawalls, pipelines, and reservoirs. While we need both gray and green infrastructure, there are increased calls to integrate more natural landscapes and surfaces with greater permeability into our communities. Often considered as a cost-effective complement to gray infrastructure, green infrastructure is going to play an even more important role as we prepare to experience more frequent and intense precipitation extremes in many parts of the United States and around the world.

We can advocate for green infrastructure in our communities and also deploy it in our own yards, gardens, and rooftops, with options like rain gardens.

▸ **Will you go green with the infrastructure at your home?**

PERMEABLE PAVEMENT COSTS 20% LESS UP FRONT THAN CONVENTIONAL PAVEMENT SYSTEMS AND IS CHEAPER TO MAINTAIN.

GREEN ROOFS ARE UP TO 40 DEGREES COOLER THAN CONVENTIONAL ROOFS.

RAINWATER HARVESTING CAN FULFILL 21%–90% OF A CITY'S ANNUAL NON-POTABLE WATER NEEDS.

GREEN INFRA-STRUCTURE BY THE NUMBERS

BIOSWALES FILTER OUT AS MUCH AS 80% OF TRACE METALS, OIL, AND GREASE FROM STORMWATER.

URBAN TREE CANOPIES CAN CAPTURE AS MUCH AS 4,000 GALLONS OF RAIN A YEAR.

RAIN GARDENS ARE 30% MORE ABSORBENT THAN A CONVENTIONAL LAWN.

DOWNSPOUT DISCONNECTION OF 26,000 PROPERTIES FROM THE SEWER SYSTEM IN PORTLAND, OREGON DIVERTED 1.2 BILLION GALLONS OF STORMWATER FROM THE SEWER SYSTEM OVER THE COURSE OF 18 YEARS.

ACTION 72

PLANT TREES TO SHADE HOUSES AND BUILDINGS

Extreme heat events are not only getting more frequent and intense in many regions, but they often cover a bigger area than in the past. Urban environments can further amplify extreme heat, with some urban areas like New York City recording average air temperatures around 5°F (2.8°C) hotter than surrounding areas. In the Seattle area, a recent study mapping heat around the community noted up to a 20°F (11.7°C) difference between urbanized and natural landscapes. This is referred to as the heat island effect. Why such a temperature difference? Urban environments tend to be warmer than surrounding areas because buildings and road surfaces absorb heat, less tree cover and shade means there are fewer cool areas, and the HVAC systems that cool buildings produce extra heat.

With extreme heat as a primary threat to human health and well-being, protecting our communities from these impacts is key. While there are numerous things we can do as individuals to protect ourselves and families during heat events (see Action 73), community-level response to extreme heat, including efforts to reduce the urban heat island effect and help protect people in our communities, is imperative. Fortunately, there are some straightforward, cost-effective solutions to the urban heat island effect. One natural solution is to increase the amount of shade and natural landscapes in our communities. Planting trees brings benefits like sequestering carbon, but their green canopies offer shade and improvements in air quality, and help to cool buildings. The strategic planting of trees, vines, and other vegetation to cover parking lots and the sides of buildings has shown to be an effective action for reducing the heat island effect. Planting these trees also makes good dollars and cents: research shows that every dollar spent on urban tree planting provides a return on investment of anywhere between $1.50 to $3.00.

Many communities are working on strategic tree-planting programs to help alleviate the urban heat island effect. Increasingly, part of this strategy involves responding to historical decisions, including racist housing practices, that have led to communities

TREES AND VEGETATION LOWER SURFACE AND AIR TEMPERATURES BY PROVIDING SHADE AND THROUGH EVAPOTRANSPIRATION. SHADED SURFACES MAY BE 20° TO 45°F COOLER THAN THE PEAK TEMPERATURES OF UNSHADED MATERIALS.

—US Environmental Protection Agency

of color and low-income communities being more exposed disproportionately to the impacts of extreme heat. As we look to take action, we need to first address the areas in greatest need across a range of community services and urban planning efforts. This includes tree planting, the location of cool air shelters, open space conservation, prioritization of green and gray infrastructure design (see Action 71), and deployment of energy-efficient building upgrades and retrofitting programs.

▶ **Will you help your community plant more trees to shade buildings, homes, and green spaces where they are most needed?**

HEALTH AND WELL-BEING

CAH Weather | **EXTREME TEMPERATURES**

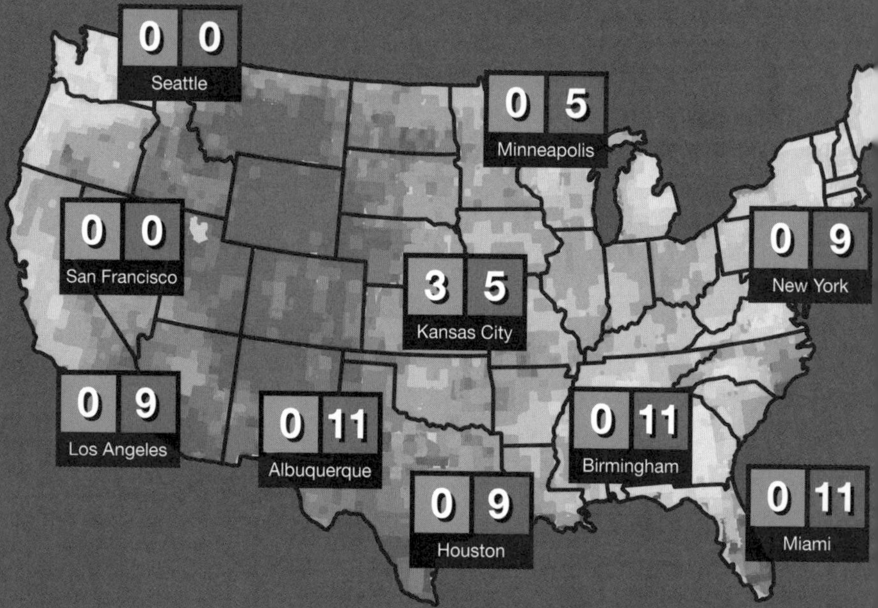

Projected Deaths Due to Extreme Temperatures

Combined Mortality Rate (Deaths per 100,000 Residents)

2 — 2000 Baseline; 2 — 2100 Projection (without mitigation)

City	2000 Baseline	2100 Projection
Albuquerque	0–2	11–14
Atlanta	0–2	9–10
Birmingham	0–2	11–14
Boston	0–2	3–4
Chicago	0–2	5–8
Cincinnati	0–2	5–8
Denver	0–2	0–2
Houston	0–2	9–10
Kansas City	3–4	5–8
Los Angeles	0–2	9–10
Miami	0–2	11–14
Minneapolis	0–2	5–8
Nashville	0–2	9–10
New Orleans	0–2	15–18
New York	0–2	9–10
Philadelphia	0–2	9–10
Phoenix	0–2	5–8
Salt Lake City	0–2	5–8
San Francisco	0–2	0–2
Seattle	0–2	0–2
St. Louis	0–2	5–8

*Extremely hot days are defined as those with a daily minimum temperature warmer than 99 percent of the days in the period 1989–2000. Extremely cold days are defined as those with a daily maximum temperature colder than 99 percent of the days in the period 1989–2000.

ACTION 73

PROTECT YOURSELF AND YOUR COMMUNITY FROM EXTREME HEAT

In a warming world, the negative impacts of extreme heat are expected to worsen. Not only have heat waves become up to two times larger in area than in the 1980s, they occur roughly three times more frequently than they did 60 years ago. It is virtually certain that heat extremes have increased in frequency and extent globally, and human-induced climate change is the key driver of this change. In June 2021, the Pacific Northwest experienced an unprecedented extreme heat wave. This event was deemed to have been impossible without human-caused climate change, and it led to serious ecosystem damage, infrastructure failures, and hundreds of deaths across Washington, Oregon, and British Columbia, Canada. With continued global warming, events like this in the Pacific Northwest are expected to occur about every 5 to 10 years!

In the United States, heat is the number one weather-related cause of death, and results in the loss of tens of millions of otherwise productive work hours. By 2100, under high emissions of greenhouse gases, it is expected that there will be an average 97,000 heat-related premature deaths each year in the contiguous United States. Even with moderate reductions in greenhouse gas emissions, that number would be 36,000—better than with high emissions but still three times higher than the current annual average. Most susceptible are people over 65 years of age, children, those with cardiovascular or respiratory illness, and those living in lower-income communities.

The good news? With preparation and comprehensive heat response plans, deaths associated with heat can be lowered. Know how to protect yourself from heat exhaustion and heat stroke. Make sure you know where to find cool air if you don't have it at your home or office, be it in a public space, library, or community cool air shelter. Avoid strenuous activity, wear loose-fitting clothing, take a cool bath, and stay hydrated. Be community minded—check on your neighbors, particularly those who might be more susceptible to heat, to ensure they are safe or provided with any needed medical attention or other services.

▸ **Are you and your community prepared for extreme heat?**

ACTION 74

PROTECT YOUR AIR

Actions and policies that reduce the combustion of fossil fuels can lead to rapid improvements in air quality and help prevent premature deaths. An estimated 17,000 to 20,000 deaths per year are attributable to air pollution from transportation, particularly the fine particulate matter called PM2.5. PM2.5 is associated with a range of health issues, including lung and heart problems. Shockingly, 90 percent or more of children around the world breathe air with particulate matter levels that exceed World Health Organization guidelines. Despite these eye-popping numbers, recent research suggests that US regulations of vehicle emissions since 2008 have already had a measurable positive impact, with deaths attributable to emissions of particulate matter and greenhouse gases dropping from 27,700 in 2008 to 19,800 in 2017. In addition to lives saved, this same research suggests that these emissions reductions came with $270 billion in economic benefits.

While there is no question that this counts as a win, researchers are careful to point out that most of these gains are associated with reductions in pollutants like particulate matter rather than carbon dioxide, a key greenhouse gas driving climate change. With more of us driving larger cars and traveling farther, even with greater fuel efficiency and "cleaner" cars, transportation greenhouse gas emissions have actually gone up in recent years. As we look to improve the health of our air, we need to continue to regulate particulate matter, ozone, sulfur dioxide, carbon dioxide, and nitrogen dioxide—all of which can impact our health and well-being.

Reductions in air quality don't just come from cars and industrial activities. Smoke from wildfires, which are growing larger in a warmer world, is linked to estimated tens to hundreds of thousands of premature deaths annually across the globe. Climate change also spells worse news for people with allergies, with grass pollen production ramping up and allergy seasons getting longer.

It is critical to note that evidence clearly shows that the impacts of poor air quality are not equal. In the United States, race is an important factor determining one's exposure to poor air quality—a 2021 EPA-funded study found that communities of color including African Americans, Hispanics, Asians, and other people of color are disproportionately exposed to fine particulate matter like PM2.5.

What can you do? Contribute less to air pollution by carefully considering what car and how far you drive, and advocating for clean air policies locally, nationally, and globally. If you have children in your life, learn about the air quality at their school and consider engaging your school in an air quality monitoring program like PurpleAir. PurpleAir sensors measure air quality in homes, schools, and communities and transmit the data in real time to a map that allows anyone to monitor their local air quality. If you are living through a wildfire smoke event or other hazardous air quality event, protect your indoor air quality by filtering your air. You can buy a room air purifier, or you can simply add a HEPA filter to the back of a box fan. An internet search for "DIY air filter" will yield a variety of videos and how-tos.

▸ **What steps will you take to protect the air you breathe?**

90% OR MORE OF CHILDREN AROUND THE WORLD BREATHE AIR WITH PARTICULATE MATTER LEVELS THAT EXCEED WORLD HEALTH ORGANIZATION GUIDELINES.

ACTION 75

PREPARE FOR MORE PESTS

According to the US Centers for Disease Control and Prevention, between 2004 and 2016, disease cases from mosquito, tick, and flea bites tripled! When it comes to these insect pests, the climate change connections are clear. Ticks, for example, which thrive when it is warm, are benefiting from shorter winters. Thanks to warming climate conditions, ticks are also emerging earlier in the season and expanding into new areas, including parts of Canada, where they were previously unable to survive. This has resulted in more ticks and tick-borne illnesses. In fact, between 1991 and 2019, Lyme disease cases in the United States nearly doubled, with the uptick partially attributed to climate change.

Experts suggest that by 2050, nearly half the world's population will be exposed to disease-spreading mosquitos. The Aedes aegypti mosquito, a culprit for spreading dengue, chikungunya, and Zika virus, is expected as far north as Chicago! Research suggests that the range of the Aedes aegypti is moving northward 155 miles each year and that the number of mosquito-friendly days is increasing as the climate warms. In addition to climate change, human migration and travel will continue to contribute to the spread of mosquitos.

So what can you do? Our climate action portfolios can include being prepared for these climate and environmental changes. Cover up with light-colored clothing, use an EPA-registered insect repellent, consider insect netting on strollers when outdoors for prolonged periods of time, and conduct regular tick checks after being outside. Try to keep standing water at a minimum outside of your home, and seek medical attention if you think you've been bitten by a tick. You can search online for telltale signs of tick bites and advice for how to successfully remove a tick and test it for disease.

▸ **How will you prepare for more pests?**

DAYS WITH MOSQUITOES

IN THE SAN FRANCISCO, CA AREA

1980s

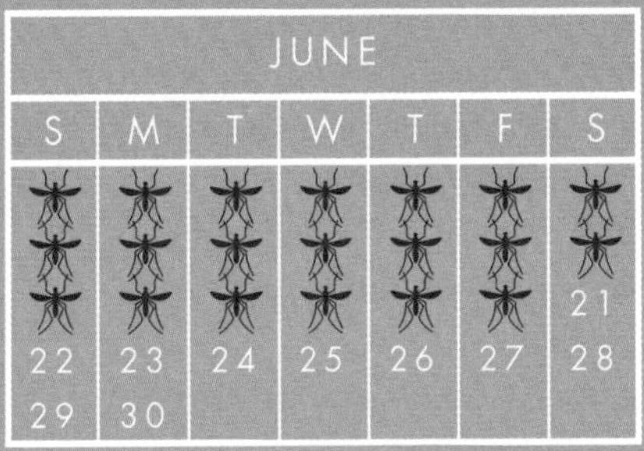

JUNE
S M T W T F S
21
22 23 24 25 26 27 28
29 30

182

DAYS PER YEAR

2010s

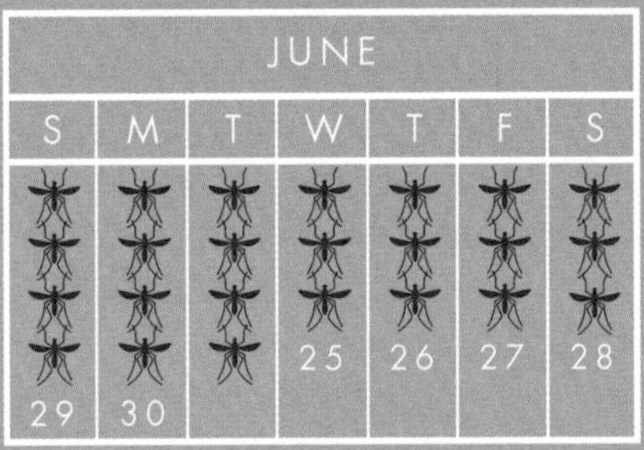

217

DAYS PER YEAR

+4 DAYS PER MONTH AVERAGE
DURING MOSQUITO SEASON
COMPARED TO 1980s

AN UPTICK IN LYME DISEASE CASES IN THE UNITED STATES

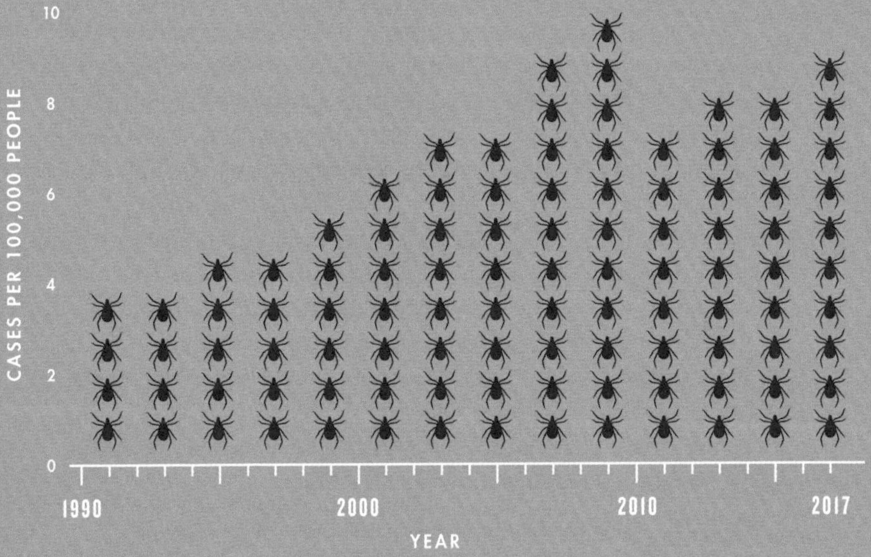

ACTION 76

DON'T DISCOUNT CLIMATE ANXIETY

Of the majority of those in the United States who agree that global warming is happening, 63 percent report that climate change is already having an impact on their health, with 48 percent reporting that climate change is affecting their mental health. Climate change is also causing "eco-anxiety" in youth, with 60 percent of youth surveyed saying they feel very or extremely worried about climate change, choosing words like afraid, sad, anxious, and powerless to describe their emotions about climate change.

The direct impacts of a warming world can include the trauma and stress that stem from living through an extreme event and experiencing property damage, injury, or loss of livelihood. Indirect impacts include anxiety related to the profound changes we may hear about in the news or online, as well as a general sense of despair around inaction by governments and others who can help to solve the problem at scale.

Solastalgia, a term increasingly used in relationship to lived experiences and the losses people associate with climate change, refers to "the homesickness you have when you are still at home." Solastalgia is increasingly being used to refer to environmentally induced distress. Feelings associated with the loss of your favorite winter activities due to climate impacts in your community are forms of solastalgia.

It is critical to care for our mental health as we live through and observe the very real toll of climate change. Individual self-care—like seeking help from a mental health practitioner, engaging in climate actions, or participating in arts activities—can help some feel less stress or hopelessness. Engaging in disaster planning within your community can help you feel more prepared and connected, and strong social ties can help communities better weather the storm. We can help children and young people by talking with them and acknowledging the range of emotions we may feel, such as fear, loss, and anger. Spending time in nature, and helping to ensure ample green spaces are available to all members of our community, can help reduce stress and stress-related illnesses. We also need to advocate for the expansion of currently under-resourced mental health care services.

▸ **Will you be mindful of your head and heart as part of your climate work?**

BE

MINDFUL

OF BOTH YOUR

HEAD

AND

HEART

AS YOU

CONTINUE

ON YOUR

CLIMATE

JOURNEY.

ACTION 77

EXPRESS YOURSELF CREATIVELY

Visual art, music, writing, poetry, theater, and other forms of creative expression are important tools in our climate action toolboxes. Creative expression can help people connect with and engage in complex topics, including climate change. The arts can be emotional and inspirational, can provide a new perspective for looking at a problem or the world, and can also be associated with creating a sense of responsibility and solidarity. While research suggests that some art may be more effective than others for inspiring and connecting people to solutions, all art can have personal and social benefits. The very act of creation, through a hobby, crafting, or other creative pursuit, is also beneficial to our mental health.

As we engage in the complex topic of climate change, harnessing your creativity and expressing yourself through the arts can be a powerful tool. If you don't feel crafty or creative, look around for climate art and artists. You might be surprised by what (and who) you discover. The climate stripes, displayed here, showing average global temperature from 1850 (cool colors) to 2021 (hot colors), were created by Professor Ed Hawkins at the University of Reading. These stripes can now be found on scarves, t-shirts, websites, book covers, and more! Seattle-based artist Anna McKee traveled to Antarctica and Greenland with researchers to help tell the stories of change in these icy environments. A range of art exhibits around the world are increasingly featuring climate and environmental art. Many musicians, like Jaden Smith and Billie Eilish, are also using their craft to elevate climate change while projects like the ClimateMusic Project are creating music from historical and projected climate data.

▸ **What kind of creative expression will you create and seek out?**

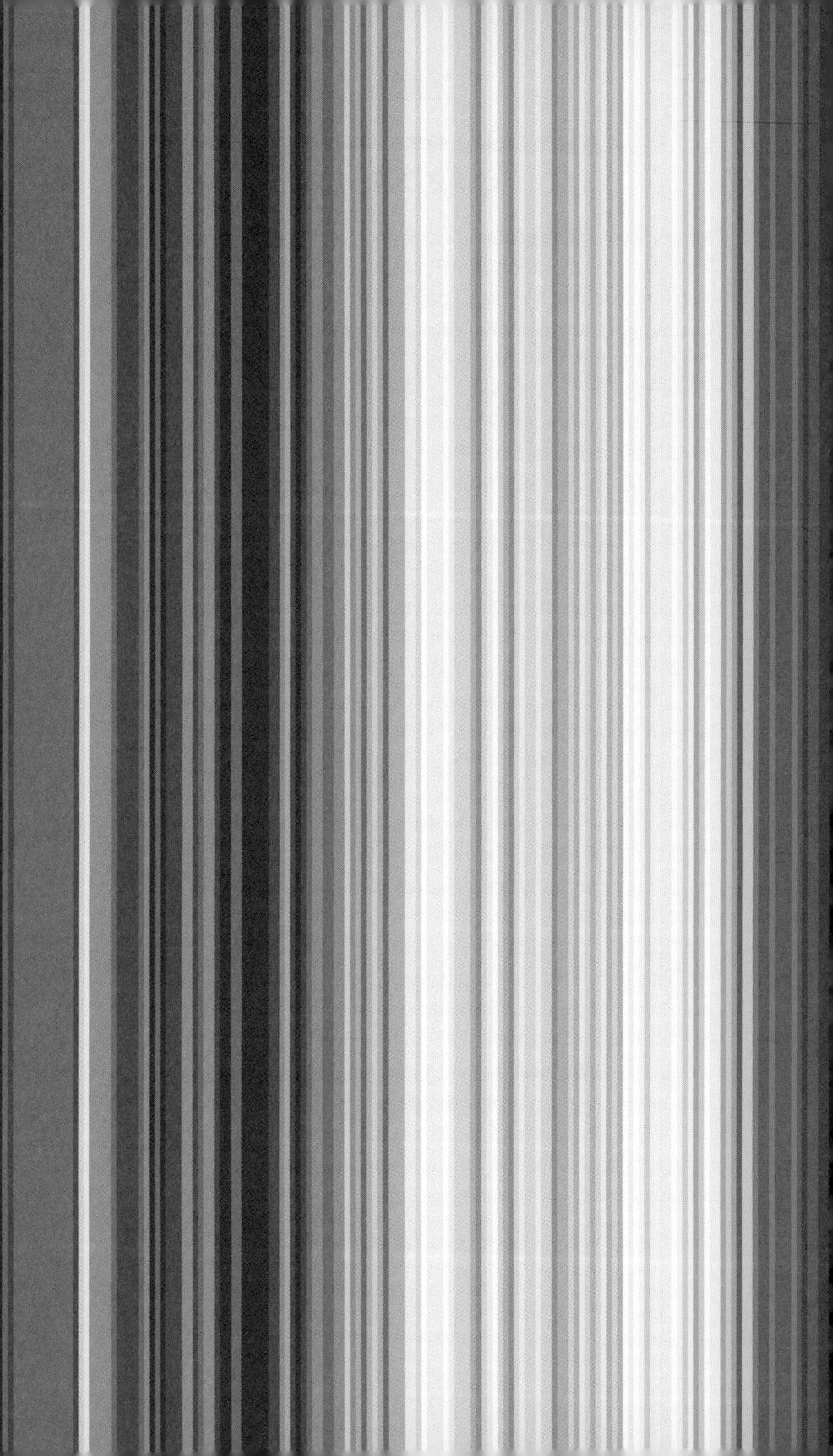

ACTION 78

BUY BEAUTY PRODUCTS RESPONSIBLY

Did you know that personal care products like shampoos, perfumes, deodorants, and lotions contribute to air pollution? One 2018 study even documented a spike in petroleum-based chemical compounds common in many personal care products in the air during morning commuting times! While plumes from our personal care products following our morning showers and makeup routines are not the same as the fumes generated by the fuel in our cars, all of these emissions are the building blocks of smog that can reduce air quality. In fact, this same study found that 40 percent of chemicals added to consumer products end up being emitted into the atmosphere.

With a reported 120 billion units of packaging used annually for cosmetics alone, along with the petroleum-based compounds used in these products and packaging, and the fuel used for transportation, there is clear room for improvement across beauty brands and personal care products. Product lines and formulations may also need to change, with rising temperatures leading to more damage to cosmetic products sensitive to the heat. Who wants melted lipstick or curdled face creams? Research suggests that even opting for "green" products can come with unexpected negative impacts. For example, the website Beat the Microbead tracks a large database of personal care products for the presence of microplastics and other potentially harmful ingredients. You might be surprised by what you find.

Supporting brands that are working to make their entire business better for the climate and environment is key, as is looking for less packaging and packaging that can be reused or recycled. More and more beauty brands are making climate commitments and increasing investments in sustainability across their businesses. As with most corporate commitments, be sure to watch out for greenwashing techniques from personal care brands, and always try to use less of the products you already have. Remember, in many cases, a little can go a long way!

▸ **Will you do your part to support climate-friendly beauty brands?**

PERSONAL CARE PRODUCTS CAN POLLUTE.

40% of chemicals added to consumer products end up being emitted into the atmosphere.

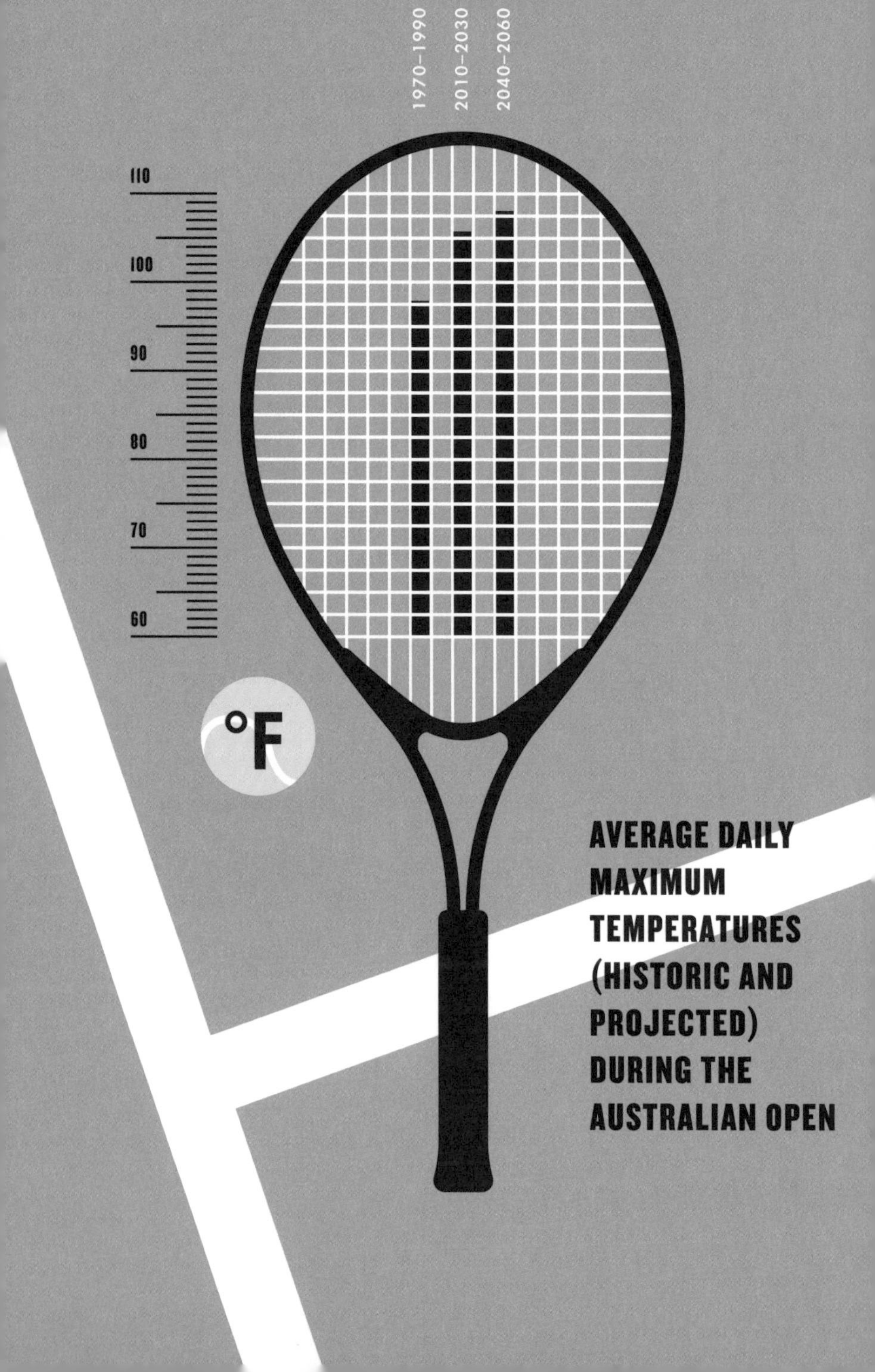
1970–1990
2010–2030
2040–2060
110
100
90
80
70
60
°F
AVERAGE DAILY MAXIMUM TEMPERATURES (HISTORIC AND PROJECTED) DURING THE AUSTRALIAN OPEN

ACTION 79

CHANGE YOUR FITNESS PATTERNS

As the world warms and we navigate through a range of changing extremes, there is both good and bad news when it comes to our fitness habits. While current warming may have led to increased levels of outdoor exercise, these gains are expected to be short-lived, especially in areas nearer to the tropics. Our ability to safely exercise outdoors in a warming world is a bit of a Goldilocks situation. Too hot and levels of exercise tend to decline along with a decreasing likelihood that amateur and professional athletes will beat personal records. Too cold and exercise levels tend to decline. Geographic regions that experience warmer winters are likely to see a seasonal increase in fitness activity, also while those who experience hotter summers will trend toward seasonal decline.

Overall, there will probably be fewer days that are safe or suitable for our regular exercise habits—due to reductions in air quality, more extreme heat days, flooding, sea-level rise impacting sports fields, and shifting seasons. All of these changes mean there may be fewer days that are safe or suitable for a range of our regular outdoor exercise habits. This includes an estimated decrease in days safe for the practice of outdoor team sports. As a concrete example, following poor air quality alerts between 2003 and 2010, "vigorous" physical activity was reduced by 18 percent in the United States while similar alerts between 2008 and 2013 in Australia led to a 15 to 35 percent reduction in cycling activity.

What can you do? Consider cross-training—embracing multiple sports across seasons, including some indoors—which can up your odds of good health and consistent exercise. Shifting the time of day when you exercise, planning your clothing carefully, and bringing sufficient water as well as adjusting your level of activity can help keep you healthy and safe during periods of extreme heat or high humidity. You may even discover that mixing up your routine inspires you to exercise more!

▸ **What activity will you add to your exercise routine?**

ACTION 80

CHERISH WINTER RECREATION

While winter is sometimes used as a scapegoat for climate-change deniers, winter is the canary in the coal mine for climate change. Places like Minnesota are actually experiencing their most dramatic warming in the winter months. Minnesota's average daily winter low temperature has warmed over 15 times faster than its average daily summer high temperature. This doesn't mean cold weather is going away, but the average characteristics of winter are changing.

Warmer winters can stress water supplies, support the survival of pests that need frigid temperatures to keep their populations in check, and can negatively affect winter recreation. In the United States, tens of millions of people engage in winter sports like skiing, snowboarding, and snowmobiling. From 2015 to 2016, winter recreation brought $20.3 billion in economic value to the US economy and supported nearly 200,000 jobs. Activities that rely on snow—including glacier tourism and skiing—and the many cultures, identities, and subsistence practices connected to winter, are already experiencing negative impacts from loss of snow cover and loss of winter.

What can we do? The winter recreation industry needs to reduce emissions from operations like snowmaking, which is both water- and energy-intensive. This shift needs to happen even as many ski resorts are relying on snowmaking to stay open. Some winter enthusiasts are changing their habits as snow gets harder to find. Some downhill skiers are swapping out the chairlift for a sweaty climb up a mountain as they seek higher elevations with snow. You can also join professional and amateur winter athletes alike in channeling your energy toward climate education and advocacy. In 2007, a pro snowboarder founded the organization Protect Our Winters (POW); today, it is 130,000 people strong. Even if winter isn't your favorite season, finding a group of people who share similar goals and interests can help drive climate action. This shift from "I" to "we" can help sustain climate work and lessen feelings of helplessness and hopelessness.

▸ **Will you cherish your wintertime activities and consider advocating to protect the seasons and activities you love?**

DAYS WITH BELOW-FREEZING TEMPS EACH YEAR IN COLORADO

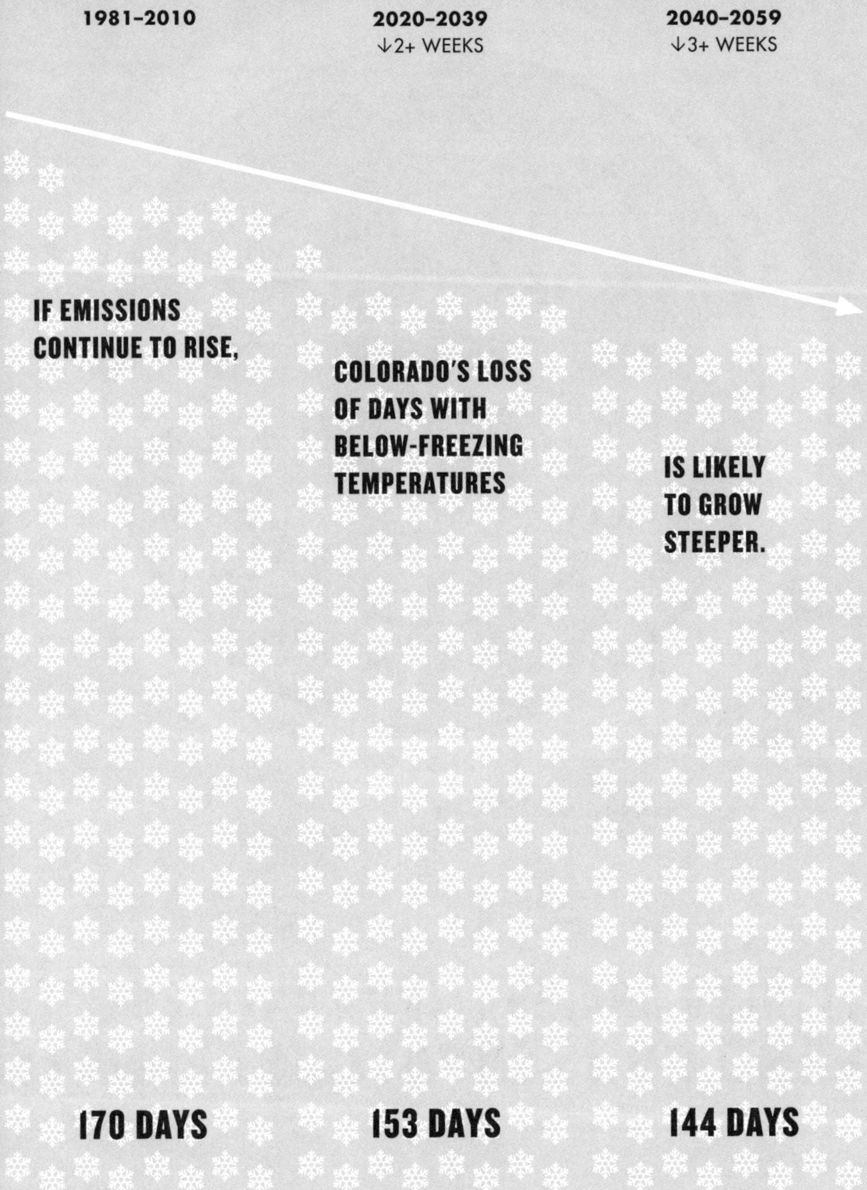

FISHING AND HUNTING CONTRIBUTE TO LAND CONSERVATION EFFORTS

ACTION 81

PAY LAND USE FEES

Hunting, fishing, and wildlife watching are growing in popularity. In 2016, 40 percent of Americans 16 years or older—that's over 101 million people—engaged in wildlife-related activities. Shifts in climate are bringing changes to the amount and location of suitable habitat, migration patterns, and ocean, lake, and stream temperatures, which means there will be impacts on the outdoor recreation activities beloved by so many.

In 2018, the United States' Fourth National Climate Assessment noted that the ranges of land-based animals have moved at a rate of 3.8 miles per decade, and the ranges of some marine animals have moved at a rate of over 17 miles per decade. As these ranges keep shifting, we can expect to see changes in the type, quantity, and location of species available for hunting or angling. Extreme heat alone is projected to result in a 15 percent decline in fishing along ocean shorelines. Big game is expected to move northward and to higher elevations; fish are expected to move to find more suitable habitat, with many species experiencing population decline; and pests and disease are expected to stress animals like moose.

Even if hunting and fishing don't align with your interests and beliefs, be aware that these activities contribute sizably to local economies and land conservation efforts, and are important to cultural heritage and traditional lifeways, in many parts of the United States. Licenses and permits represent a big portion of the dollars that the US government uses to manage and conserve these public spaces. In 2021, the US Fish and Wildlife Service reported that hunting license sales totaled more than $918 million. Over $750 million was spent on fishing licenses the same year. Federal excise taxes—on things like ammunition, guns, trolling motors, and fish tackle—have generated billions of dollars in additional funds that support state-level conservation and recreation projects. The Department of the Interior has also disbursed over $20 billion to states to advance their conservation efforts.

This revenue is critical, so if you use public lands for outdoor recreation—be it birdwatching, hunting, or simply basking in nature while on a camping trip or hike—pay use fees and purchase all required permits, tags, or licenses. Future generations will thank you.

▶ **Will you commit to paying the required fees for your outdoor recreation?**

CIVIC AND COMMUNITY ENGAGEMENT

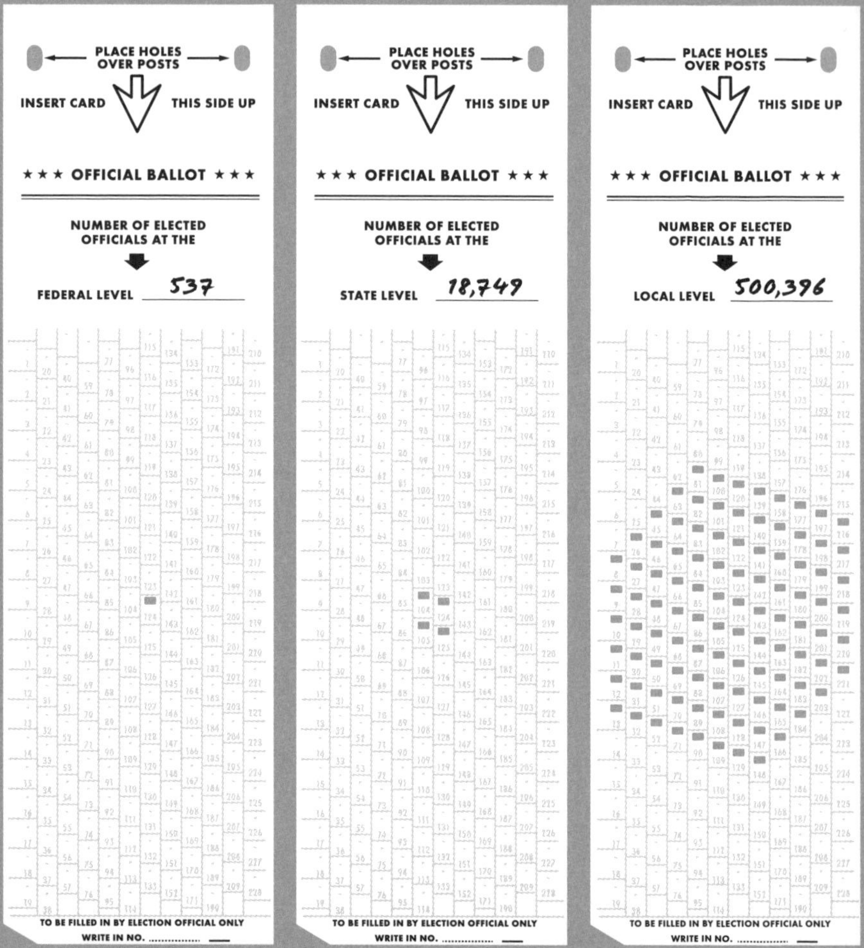

0.1%

OF ELECTED OFFICIALS ARE AT THE FEDERAL LEVEL.

3.6%

OF ELECTED OFFICIALS ARE AT THE STATE LEVEL.

96.3%

OF ELECTED OFFICIALS ARE AT THE LOCAL LEVEL.

THINK GLOBALLY. VOTE LOCALLY.

ACTION 82

VOTE IN EVERY ELECTION

When many in the United States think about voting or political action, we generally think about the president or senators and maybe our state's governor, but did you know that there are more than 519,000 elected officials across all levels of government in the United States? The 2017 US Census of Governments tallied 90,126 government units in the United States, including federal, state, county, township, and municipal governments. That's 90,126 places where climate action can take place—from school districts to statehouses to the three branches of the federal government.

While a majority of attention and concern is directed at our federal and state-level elected officials, local officials account for 96 percent of our elected leadership! These local leaders like city elected officials, county commissioners and officers, and school board members are responsible for distributing nearly $2 trillion for local services and programs. They can also play a critical role in advancing climate change–related investments, priorities, and regulations in our communities. Each one of these decision-makers should be helping to address climate change in our communities.

Importantly, these positions are often elected, allowing us to actively vote for the people who will make decisions on our behalf. Yet 73 percent of eligible voters in the United States don't vote in local elections. We know that the voters casting ballots in local elections tend to be white, affluent, and 65 years or older. This means people who are least likely to live through the worst of climate change are deciding and influencing both today's and tomorrow's priorities.

What can you do? If you are eligible to vote, register. Then vote. In every single election.

▸ **Will you cast your ballot in every election?**

ACTION 83

ENGAGE YOUR ELECTED OFFICIALS

Many of us are accustomed to the politicization of climate change, which frames it as a blue, liberal issue, not a red, conservative one. While it is true that climate change and global warming remain topics of partisan debate, there is a growing mismatch between the priorities and positions of our elected officials and the desires and wants of their constituencies.

Across Democrats, Republicans, and independent political affiliations, registered voters in the United States want elected officials to do more to advance climate-related policies. Polling in September 2021 indicated that a majority of registered voters want to see more climate action, with 60 percent saying "global warming should be a high priority or very high priority for the president and Congress." Most registered voters want more climate-friendly energy policies to be adopted, including scaling up renewable energy such as wind and solar, incentivizing energy-efficiency upgrades to homes and businesses, and regulating carbon dioxide as a pollutant. While this polling does show a different sense of urgency and responsibility across political affiliations, the main takeaway is that a significant majority of registered voters want to see more action by our federal government and local leaders.

How can we close the chasm between voter concern and action by our elected leaders? Email, call, or attend a meeting, and urge your elected officials to do more. Your call to action can lead to the adoption of policies and incentives that develop and scale clean energy, create tax incentives and rebates for increasing energy efficiency of homes and businesses, and help to conserve land and waterways. Your urging can encourage leaders to support communities that are disproportionately harmed by air and water pollution, and to make investments in community-based climate adaptation and resilience measures. If you're not sure where to start, check out USA.gov/Elected-Officials for more information.

▸ **Will you pick up the phone, email, and engage in a conversation about climate change with your elected leaders?**

A

MAJORITY

OF REGISTERED VOTERS IN THE US WANT TO SEE

MORE CLIMATE ACTION,

WITH 60% SAYING GLOBAL WARMING SHOULD BE A HIGH PRIORITY OR

VERY HIGH PRIORITY

FOR THE PRESIDENT AND CONGRESS.

ACTION 84

CHAMPION CLIMATE PLANNING IN YOUR COMMUNITY

Climate action plans—which can come in a variety of forms and levels of detail—are increasingly being developed around the United States, with a reported 600 local governments having developed such plans since 1991. In addition, many Tribes across the United States have climate plans in place. Thirty-three states have developed climate action plans or are in the development or revision process. These plans provide a framework or road map for measuring and assessing locally specific climate-related risks, articulating current greenhouse gas emissions from different activities, and outlining opportunities for, and commitments to, reducing those emissions. They can include strategies and priorities intended to reduce a region's exposure and increase its resilience to climate-related events and impacts.

Engaging in local climate planning is a great way to help shape the future of your community and ensure that climate planning includes the priorities and needs you envision. Many of these plan development processes are starting to include more comprehensive community engagement opportunities and listening sessions, and often are pulled together with direct support from local residents and volunteers.

The first step is to see if your local, county, and state governments have plans in place. If so, take a look at the plans to assess how they are being implemented and whether they are on track. A 2020 Brookings Institution report noted that while about half of large cities in the United States have greenhouse gas reduction targets established, a majority of these cities are currently lagging behind these targets. If your community doesn't yet have a plan in place, start engaging on social media platforms, listservs, and by phone to bring your community together to encourage your local and state leaders to get a planning process underway.

We know that the risks and impacts of climate change are diverse and there are few one-size-fits-all solutions. Ensuring that emissions reductions targets and community resilience and adaptation priorities are in place requires community-level planning and engagement.

▸ **Will you make sure climate plans are in place in your community and help put them into action?**

STATE CLIMATE ACTION PLANS

AS OF AUGUST 2022

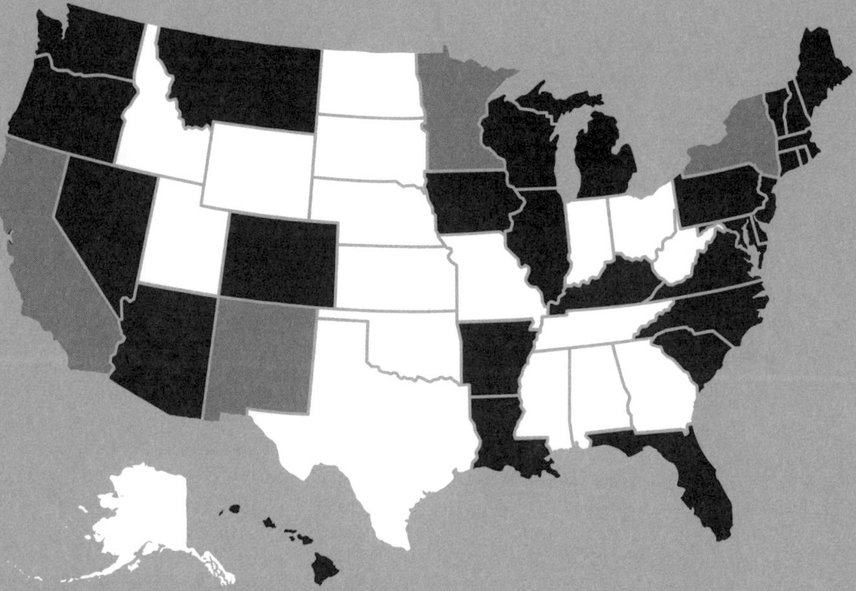

29

STATES HAVE RELEASED PLANS

4

STATES ARE UPDATING OR DEVELOPING PLANS

600

LOCAL GOVERNMENTS HAVE DEVELOPED PLANS SINCE 1991

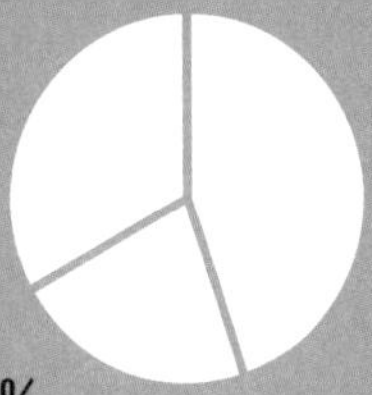

STATED EMISSIONS REDUCTION COMMITMENTS BY THE 100 LARGEST CITIES IN THE US (2017)

40%

OF THE US POPULATION LIVES IN BIGGER CITIES WITH ACTIVE CLIMATE ACTION PLANS

2/3

OF CITIES ARE NOT ACHIEVING THEIR TARGETED EMISSIONS LEVELS

CITY WITH LARGEST DECREASE IN EMISSIONS COMPARED TO 1990 BASELINE LEVELS:

-47%

LOS ANGELES, CA

CITY WITH LARGEST INCREASE IN EMISSIONS COMPARED TO 1990 BASELINE LEVELS:

+39%

TUCSON, AZ

ACTION 85

CONTRIBUTE TO COMMUNITY GROUPS

Many of the climate actions we hear about often focus on reducing our personal emissions and changing other individual behaviors. This book includes plenty of those examples, but giving the gift of your time, money, or services can also be a meaningful climate action.

There are a lot of organizations that can use your support. Numerous community groups, as well as local, regional, national, and international nonprofits, have aspects of climate mitigation or adaptation and resilience at the heart of their work. Some of their activities include:

- Providing disaster response and recovery services
- Expanding access to physical and mental health care services
- Increasing outdoor recreation opportunities
- Reducing greenhouse gas emissions
- Expanding access to climate news and educational materials
- Expanding green spaces, tree cover, or bike lanes and public transportation

Not sure where to start? Look for a local climate or environmental justice group, or chapters of national or international organizations focused on environmental justice and climate mitigation or adaptation. Sites like Charity Navigator or 1% for the Planet can help you locate highly rated environmental nonprofits and other groups advancing critical social and environmental justice efforts. You can also search the online fundraising site GoFundMe, which has a section dedicated to environmental and social causes of all shapes, sizes, and geographies. Still stuck? Even better, ask your community of friends and coworkers for recommendations of organizations worthy of your support.

▸ **What organizations are taking climate action in your community? How can you help to support their work?**

COMMUNITY GROUPS CAN LEAD THE WAY ON CLIMATE ACTION IN MANY WAYS

- **Providing disaster response and recovery services**
- **Expanding access to physical and mental health care services**
- **Increasing outdoor recreation opportunities**
- **Reducing greenhouse gas emissions**
- **Expanding access to climate news and educational materials**
- **Expanding green spaces, tree cover, bike lanes, and public transportation**

YOUTH CLIMATE ACTIVISTS

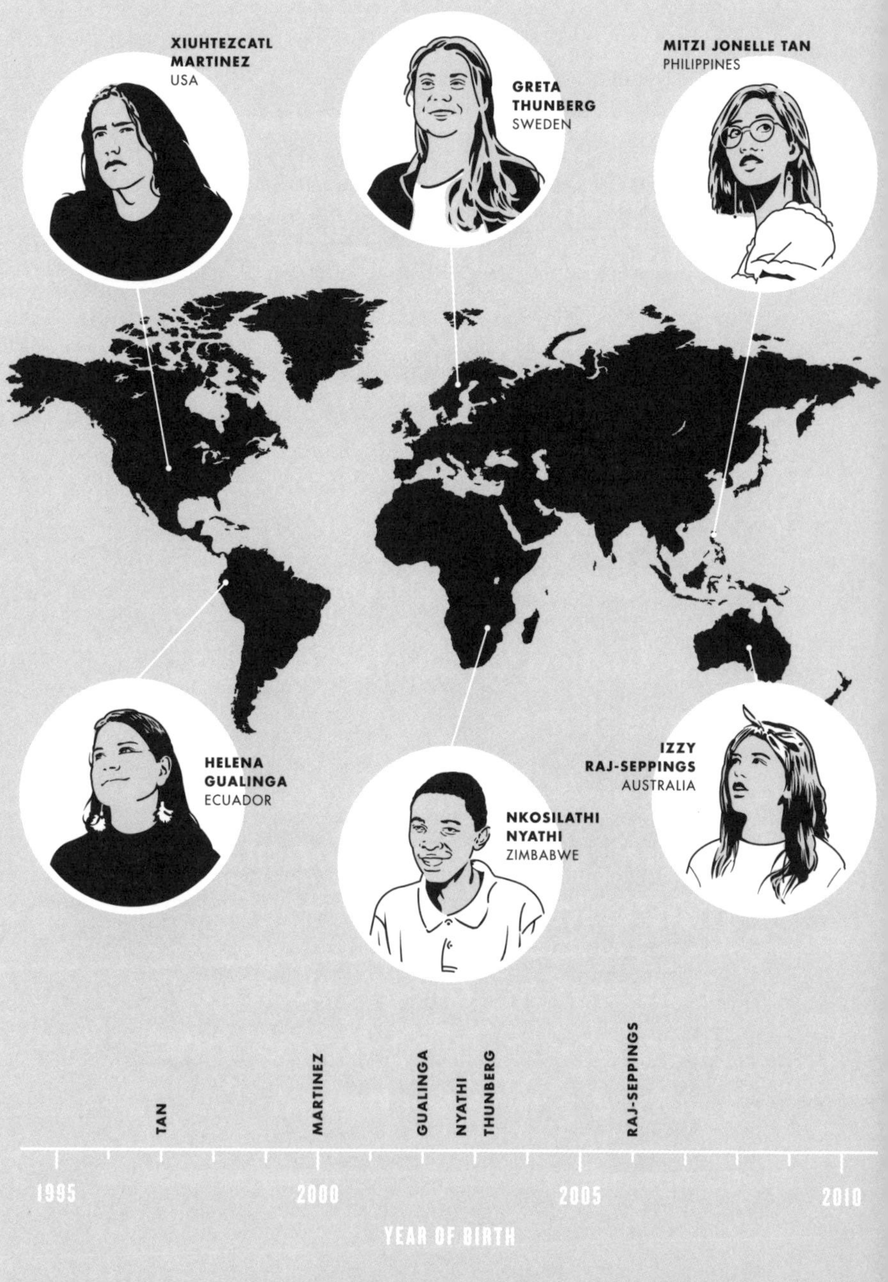

ACTION 86

SUPPORT YOUTH CLIMATE ACTIVISM

"Our hopes and dreams drown in their empty words and promises," said youth climate activist Greta Thunberg in a speech at the 2021 Youth4Climate summit in Italy. "Of course, we need constructive dialogue, but they've now had thirty years of blah blah blah, and where has that led us?" The they in her speech are global leaders who have failed to choose a different climate future for humanity, despite decades of mounting evidence of the consequences of inaction. As she so pointedly highlighted, to date, the words of those in power have led to little meaningful progress on climate change. More than half of all greenhouse gas emissions were generated after 1990. That means that adults today are linked to the emissions that have committed the planet to a climate-changed future.

As climate-concerned individuals and community members, we need to do our part to shift away from empty words and promises and act in service of our collective future. Part of this is honoring, supporting, and advancing the priorities and needs that the growing youth climate movement has articulated. Youth-led organizations, like the Sunrise Movement, are gaining influence as agents of change and, increasingly, access to the halls of power. In 2020, the UN secretary-general launched a Youth Advisory Group on Climate Change.

We can bolster the bold youth leadership that is influencing climate conversations and decision-making in a range of ways: march at their events, support their school strikes, and, critically, understand the agenda and needs of local, national, or international youth-led organizations working to motivate climate action. A quick internet search should yield a variety of options to consider.

The future of these young activists is at stake, and they have made serious progress in shifting the conversation among elected leaders across the globe.

▸ **Will you help move from "blah, blah, blah" to bold action by supporting a youth-led climate organization?**

ACTION 87

SHARE YOUR OBSERVATIONS AND EXPERIENCES

One factor that influences our perceptions of climate change and the risks it poses to us as individuals is our own direct observations and experiences—say, with extreme weather events or changes in temperature or precipitation patterns—in the places we call home. While human behaviors and risk perceptions are messy and complex (just ask a behavioral or social scientist!), these firsthand experiences can make the large scale of climate change feel more relatable and the risks more real.

How can we harness these personal observations and concerns to catalyze community action? Numerous organizations around the world are helping translate people's experiences and concerns into actionable localized data to drive better, more responsive climate policy. Some are pairing community members with researchers to advance solutions to their environmental challenges. Others are working to weave these powerful personal narratives into climate stories told from a perspective not often at the heart of our climate news. Many community science platforms are worthy of our time, observations, and stories. Some of my favorites include:

- **ISEECHANGE (ISEECHANGE.ORG)** collects observations and stories from individuals around the world to "help monitor climate and weather in real time . . . building resilience awareness, identifying trends, improving models, and co-creating solutions."
- **THE USA NATIONAL PHENOLOGY NETWORK (USANPN.ORG)** brings together volunteers of all ages and backgrounds to monitor the impacts of climate change on plants and animals across the United States.
- **THE COMMUNITY COLLABORATIVE RAIN, HAIL & SNOW NETWORK (COCORAHS.ORG)** is a volunteer network of backyard weather observers measuring and mapping precipitation across the United States.
- **THE THRIVING EARTH EXCHANGE (THRIVINGEARTHEXCHANGE.ORG)** connects communities and scientists to collaboratively tackle local climate and environmental challenges.

▸ **Will you share your climate observations and stories?**

"Community Science happens when communities and scientists do science together to advance community priorities.

'DOING SCIENCE'

includes defining questions, designing protocols, collecting and analyzing data, and using scientific knowledge in decision-making. Every community science project begins with community voices, is guided by community knowledge, and ends in community impact."

—Thriving Earth Exchange

ACTION 88

ROLE-PLAY CLIMATE SOLUTIONS

Having a hard time understanding how different climate actions and policies all add up? Have you ever wanted to see what it might be like to participate in United Nations climate change negotiations? You're in luck! A role-playing exercise called the World Climate Simulation offers a chance to explore the ifs and thens of different national and regional greenhouse gas emissions reduction pledges. Climate Interactive, the organization behind the game, also offers other simulations that allow you to look at the impact of different climate solutions, from deployment of new technologies to different agricultural practices, like changing crops and livestock production.

Not interested in a group activity? The computer simulator used for World Climate Simulation—called C-ROADS—can also be used by individuals. Climate Interactive and its partners also offer regular training and workshops, which include opportunities to engage in virtual simulations. If you get really into it, you can even become a facilitator and host your own local climate workshops. These simulations have been used in a range of contexts, from classrooms to the US Congress!

If role-playing and simulations aren't your speed, climate-related board games, video games, and other entertaining platforms allow you to try your hand at climate policy and learn about different climate change impacts. The National Aeronautics and Space Administration (NASA) even has a program—called Climate Kids—which includes online games on topics like coral bleaching and ocean currents. Given that 43 percent of US adults frequently play video games, and that 84 percent of teens in the United States have access to a game console, gaming offers a great opportunity to weave climate education and content into widely used platforms.

▸ **Will a game or two be a game changer for informing and inspiring your climate work?**

In the United States,

43% OF ADULTS FREQUENTLY PLAY VIDEO GAMES

and

84% OF TEENS HAVE ACCESS TO A GAME CONSOLE,

making gaming an ideal way to learn about

CLIMATE CHANGE

and

SOLUTIONS.

EDUCATION AND CLIMATE INFORMATION

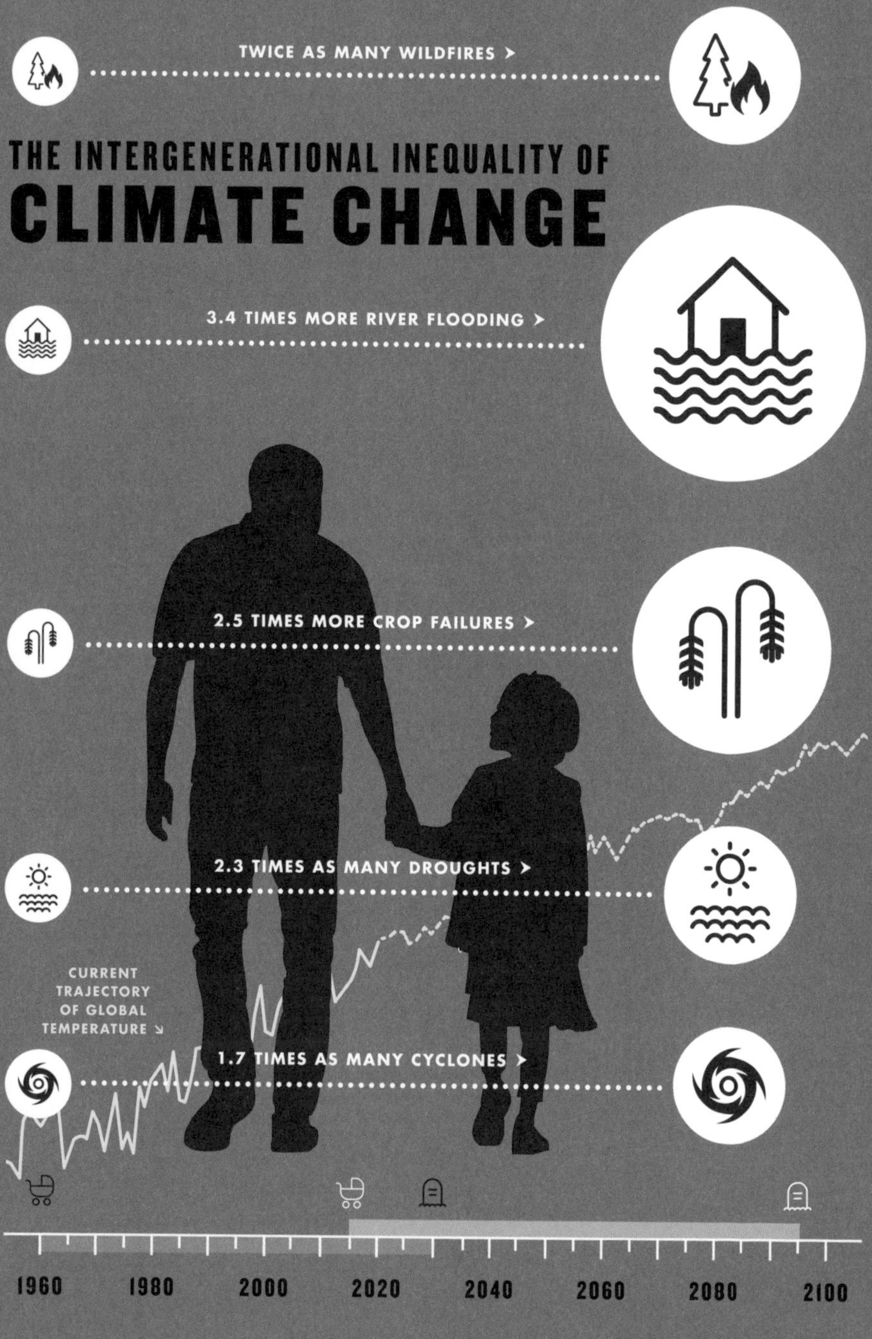

TWICE AS MANY WILDFIRES >
THE INTERGENERATIONAL INEQUALITY OF
CLIMATE CHANGE
3.4 TIMES MORE RIVER FLOODING >
2.5 TIMES MORE CROP FAILURES >
2.3 TIMES AS MANY DROUGHTS >
CURRENT TRAJECTORY OF GLOBAL TEMPERATURE
1.7 TIMES AS MANY CYCLONES >
1960
1980
2000
2020
2040
2060
2080
2100

ACTION 89

ACT ON BEHALF OF CHILDREN

If you have children in your life—and even if you don't—you might be motivated to act in service of protecting their future. Sobering research in 2021 highlighted just what's at risk for children born today: An average six-year-old will live through roughly three times more climate-related disasters than their grandparents. They are expected to experience twice as many wildfires, almost four times more river flooding, two and half times more crop failures, and over two times as many droughts as their grandparents. This is called the intergenerational inequality of climate change.

The silver lining of this grim picture is that if we limit warming to 1.5°C (2.7°F), as set out in the Paris Agreement, children born today will face dramatically reduced risks. What can parents, grandparents, and all adults do? Go all in on climate in whatever ways you can. Look at examples in this book for ways you can engage in individual and collective action. Advocate for change in your spheres of influence—whether with elected officials, faith leaders, or your own family members and friends—using your voice, concern, and expertise to ensure a safe future for all children.

Since humans are generally bad at prioritizing the future and often struggle to work in service of the collective good, we should remember that climate change is already here and touching every corner of the globe, including wherever you call home. A majority of us alive today will be living through a profoundly different climate in the next couple of decades, with even worse impacts on the horizon for younger generations. We all stand to lose, but we also all stand to gain from climate actions and investments.

The climate science is clear: we still have the opportunity to choose what climate future we will live through in the coming decades.

▸ **What motivates you to act? Will you consider your impact on future generations when deciding what actions to take?**

ACTION 90

SEEK CLIMATE SOLUTIONS FOR SCHOOLS

With an estimated 84,000 public schools and 100,000 school buildings in the United States, along with a fleet of more than 480,000 school buses, opportunities abound to make investments in schools that can bring many climate benefits. Despite the importance of our schools as a place for education and other public services (like voting and emergency shelter), state capital funding for schools fell by an estimated 31 percent, or $20 billion, between 2008 and 2017.

While overall funding is falling, America's school infrastructure is near failing, getting a D+ rating. Fifty-four percent of public school districts need updates or replacement of key building systems, including heating, ventilation, and air-conditioning (HVAC). In 2020, 36,000 schools in the United States documented needing HVAC upgrades!

What does this have to do with climate? Schools are one of the largest consumers of energy in the public sector, and energy costs are a typical school's second-highest expense. Improving efficiency of HVAC systems, reducing energy loss, and working to electrify school buses—the largest mass transit fleet in the United States—can all add up to reduce emissions and make our schools safer, healthier places to be.

What can you do? Support your local schools in getting the funding they need to tend to their infrastructure. A majority of schools report using property taxes and capital funding from state and local governments to pay for these upgrades, so engaging locally with your local school board and elected officials can help. As a parent or student, help your local schools adopt practices that reduce energy use. If you want to help tackle the problem of polluting diesel school buses, look to cities like Morris, Minnesota, where residents, partners, and allied organizations worked together to secure grant funding and other financing to buy electric school buses for their community.

▸ **Will you pass the test and help schools become healthier for people and the planet?**

PUBLIC SCHOOLS HAVE A FLEET OF OVER 480,000 SCHOOL BUSES. THAT'S OVER 480,000 OPPORTUNITIES TO

GO GREEN.

ACTION 91

TEACH CLIMATE CHANGE IN THE CLASSROOM

We know that today's children will bear the worst of the impacts of climate change in the years to come. We also know that millions of students across the United States are already missing classes due to natural disasters, which are only going to worsen with climate change. Building climate literacy in our young people is essential, yet polling in 2019 found that most teachers in the United States aren't teaching it. A 2016 study found that half of US teachers spend less than two hours across the entire year teaching climate change.

This lack of climate education in the classroom does not stem from resistance from parents, with 84 percent of parents of children under 18 years of age supportive of teaching climate change, and a whopping 78 percent of registered voters (across Republicans, Democrats, and independent affiliations) reporting wanting children to be taught "the causes, consequences, and potential solutions to global warming." Teachers also widely agree that teaching climate change is important, yet 55 percent said they're not covering it. Why? A majority felt the topic was outside of their subject area. Other barriers include concerns about parent complaints, lack of support from districts, and lack of appropriate teaching resources and materials.

What can you do as a parent or climate-concerned individual? First, actions can start at home. Have climate conversations with the children in your life. The same polling indicated that only 45 percent of parents talk about climate change with their kids (see Action 92). We can also help our teachers by ensuring sufficient funding for schools, supporting opportunities for professional development (which can include participation in a growing number of programs and teacher trainings related to climate change), and encouraging our local and state education leadership (e.g., school boards) to require teaching climate change. We can also promote programs and climate change curriculum coming from trusted entities like NASA and NOAA.

▸ **Will you help teachers get the support they need to teach climate change in the classroom?**

SHOULD CLIMATE CHANGE

46%
OF CONSERVATIVE REPUBLICANS SAY YES.
78%
OF MODERATE REPUBLICANS SAY YES.
94%
OF MODERATE DEMOCRATS SAY YES.
97%
OF LIBERAL DEMOCRATS SAY YES.

BE TAUGHT IN SCHOOL?

78% OF ALL REGISTERED VOTERS SAY YES.

ACTION 92

TALK CLIMATE CHANGE WITH YOUR KIDS

In 2019, only 45 percent of parents in the United States reported talking to their kids about climate change. This relative silence is, understandably, often associated with the fear and trepidation of having difficult conversations. However, these fears and emotions are exactly why we need to break the silence and have climate conversations with our kids. In fact, leaning into these emotions can actually result in more impactful and meaningful conversations with our children.

Of course, climate conversations go both ways. Research shows that children who are taught about climate change in the classroom (see Action 91) often bring their learning home, engage in conversations with their family, and subsequently have a measurable impact on raising the concern of their parents about the issue. This is called the pass-through effect. Research suggests that daughters have the biggest influence on their parents' attitudes about climate change, particularly with their fathers and when their parents identified as conservative!

Kids may be the primary climate influencers in their families, but adults need to be supportive of these climate conversations at home. National Public Radio talked to several experts in 2019 and came up with a list to help guide these conversations:

- Break the silence.
- Give your kids the basic facts.
- Get outdoors.
- Focus on feelings.
- Take action.
- Find hope.

As parents, and adults with kids in our lives, we can safely open the door and invite these difficult conversations. It is likely to be of mutual benefit.

▸ **Will you have a climate change conversation with the kiddos in your life?**

Break the silence.

Give your kids the basic facts.

Get outdoors.

Focus on feelings.

Take action.

Find hope.

CALLING BULL

is a performative utterance,

A SPEECH ACT

in which one publicly repudiates something objectionable.

The scope of targets is broader than bull alone.

YOU CAN CALL BULL ON BULL,

but you can also call bull on lies, treachery, trickery, or injustice.

—CallingBull.org

ACTION 93

BE A SAVVY CONSUMER OF CLIMATE INFORMATION

Lots of tricks, traps, and communication strategies are deployed to mislead or deceive us across a range of "controversial" issues, including climate change. The goal is often to cast doubt, delay action, sow confusion or division, shift blame, and avoid responsibility . . . or all of the above. While disinformation is false information created and spread with the intent to mislead, misinformation can be spread with intent or by mistake. Both can have negative consequences. Unfortunately, a big part of the burden falls to us to be savvy consumers of information and to not play a part in spreading misinformation, which can be very easy to do on social media. It requires some work and a critical eye, but it is a vital skill in an era of rampant misinformation and disinformation.

Of course, these tactics aren't new. In 1995, scientist Carl Sagan wrote a chapter in the book *The Demon-Haunted World: Science as a Candle in the Dark* called "The Fine Art of Baloney Detection." In more recent times, a wildly popular course at the University of Washington—titled Calling Bull—was developed to help college students hone this essential skill. The course instructors have made their readings, resources, and exercises available to all of us at CallingBull.org, and they've also written a book. The whole goal is to provide us with the skills to "detect and defuse" bull.

The team at Calling Bull suggests starting with a "bull inventory," listing all the 'bull' you see over the course of a week. When you're done with your inventory, review some of their case studies and tips for becoming a pro at baloney-spotting, including how to spot misleading graphs and graphics and identify the use of artificial intelligence in imagery.

▸ **Will you work to identify bull in others' climate messaging, as well as your own?**

ACTION 94

TRACK THE STATE OF THE SCIENCE

The Intergovernmental Panel on Climate Change and its assessment process is an essential source of critical global climate knowledge. Did you know that the United States and several other nations have their own climate science reports and syntheses? In the United States, there is a federally mandated program dedicated to the coordination and synthesis of global environmental change research for the nation.

The US Global Change Research Program (USGCRP) works across 13 federal agencies to coordinate and synthesize the science of climate change, its impacts on the United States, and solutions. Part of the work of the USGCRP is the National Climate Assessment process, which results in a comprehensive report on the "state of the science," including climate change drivers and trends, climate impacts, and different responses and approaches for adaptation and mitigation actions across the nation. Each assessment, produced every four years, brings together hundreds of federal and nonfederal experts from across the United States to evaluate, summarize, and communicate critical climate change knowledge for use by policymakers, community leaders, and individuals. Authors come from a variety of disciplines and perspectives, including universities, the private sector, Indigenous communities, and state and local governments. Many of these contributors volunteer their time to the creation of the National Climate Assessment. At the time of writing, the Fifth National Climate Assessment process was underway, with anticipated completion in late 2023.

To stay up to date on the latest climate science in your area without reading the whole report, here's a tip: There is a chapter dedicated to changes in each of 10 major regions in the United States (see Overview for graphic summary), so you can focus on the chapter for your region. Looking at the author list for your region's chapter can also help you identify your local climate science leaders (see Action 95).

▸ **Will you read your region's chapter in the National Climate Assessment or look at a local climate report for your region?**

THE US GLOBAL CHANGE RESEARCH PROGRAM'S

mandate is to develop and coordinate "a comprehensive and integrated United States research program which will assist the Nation and the world to understand, assess, predict, and respond to human-induced and natural processes of global change."

—US Global Change Research Program

ACTION 95

LOOK TO LOCAL CLIMATE SCIENCE LEADERS

It can be hard to wade through the increasing volume of news on climate science, especially with the continued growth of climate misinformation. We are often left wondering what it all means for us in the place we call home. Looking for and listening to what local climate science leaders have to say can offer a means of staying up to date on the most accurate research about your region. You can find these leaders at universities and community colleges, extension offices, state climate offices, nongovernmental organizations, and tribal natural resources offices, to name a few.

Many of these organizations are expanding their capacity to deliver locally relevant climate change information to the communities they work in while others, like the University of Washington Climate Impacts Group, have been helping Pacific Northwest communities understand and prepare for climate change for over 25 years! The work of these groups often includes supporting the use of climate science in climate action or adaptation plans, in the design of critical infrastructure like drinking-water facilities, and to inform habitat preservation and restoration priorities. These groups can also be effective translators of technical scientific reports. A quick internet search or a call to your local library or university can help you identify some of the most regionally relevant climate change researchers and resources. You can also sign up for these organization's listservs, follow them on social media, participate in their conventions or webinars, invite them to speak at a local event, or send a direct inquiry.

▸ **Will you seek out information from your local climate science leaders?**

"No matter what you do, or what your skills are, you can have a role in shaping that future and thinking about how we prepare for and reduce climate change."

—DR. AMY SNOVER

ACTION 96

LOOK TO COMMUNITY LEADERS

If local science leaders and science organizations can help you wade through the climate science, local community leaders and community organizations can help you see how climate challenges are being confronted in your community. If your local government is engaged in climate planning, you might be able to volunteer to help, or provide feedback during a community engagement session (see Action 84).

Further, many local nonprofits—whether or not explicitly focused on the climate or environment—are engaging in work that can help strengthen a community to better withstand the negative impacts of a changing climate. This work could span anything from how your local health and mental health nonprofits are weaving climate change education and action into their work to how community groups are incorporating climate impacts into their critical environmental justice efforts. For example, my neighborhood community foundation has recently pulled together a group focused on climate solutions in our neighborhood. This complements efforts by my city and county leaders by building focused conversations, community, and climate actions directed quite literally in my own backyard.

A sense of belonging can keep us motivated and inspired to act even in the face of a large and looming challenge. Looking to your community leaders, or becoming a local climate leader yourself, will help us to accelerate (or maybe start!) climate action that works in, and for, our communities.

▸ **Will you support your local community organizations in their efforts to address climate change? Will you consider using your skills and talents to become a community climate leader yourself?**

Alyssa Quintyne
Alaska Center
ALASKA

Gary Cuneen
Seven Generations Ahead
ILLINOIS

Julia Kumari Drapkin
ISeeChange
LOUISIANA

Johannah Blackman
A Climate to Thrive
MAINE

Troy Goodnough
Morris Office of Sustainability
MINNESOTA

James and Joyce Skeet
Covenant Pathways/Spirit Farm
NEW MEXICO

Dr. Robert Bullard
Bullard Center for Environmental
and Climate Justice
TEXAS

David Bill and
Faith Van De Putte
Midnight's Farm
WASHINGTON

Jamie Stroble
Noio Pathways
WASHINGTON

A SMALL SAMPLE OF LEADERS
MAKING A DIFFERENCE THROUGHOUT THE UNITED STATES

WHO ARE THE CLIMATE LEADERS IN YOUR COMMUNITY?

ACTION 97

TALK ABOUT CLIMATE ISSUES WITH FRIENDS AND FAMILY

With decades of climate science clearly showing the potential perils of a warming planet, it's evident that communication by climate scientists is insufficient to motivate us to act. In fact, climate scientists are some of the least effective messengers. This is where you come in.

Research indicates that not only does it matter how we talk about climate but who talks about it. Friends and family members are some of the voices we trust most on critical issues. Yet, national polling suggests that a minority of us discuss climate change on a regular basis. How can an issue be treated like it's important if we don't talk about it? How can we expect people to act and be part of the solution if they don't see the problem in the first place? A climate conversation is a climate action with impact; talking about climate change is documented to lead to greater acceptance of the science, and can eventually lead to behavior change.

Afraid of having a climate conversation? We often underestimate pro-climate views of those around us and overestimate the amount of polarization between our views and the views of others. You also don't have to come prepared with all the facts and figures or for a super in-depth conversation. You could ask your friends, for example, if they happened to read an article you saw about climate change in your region, or invite them to join you at a local climate-related event. The most important ingredient in the conversation is you—what do you care about, what are your shared values, and what do you hope to see for the future of your neighborhood, community, or aspect of life you hold dear that is threatened by climate change?

Overall, it has never been more critical to engage in climate conversations—at home, at work, and in our community.

▸ **Who will you talk to today about climate change?**

WHILE 72% OF AMERICAN ADULTS AGREE THAT GLOBAL WARMING IS HAPPENING

AND 64% OF AMERICAN ADULTS THINK GLOBAL WARMING IS AFFECTING THE WEATHER

ONLY 35% OF AMERICAN ADULTS DISCUSS GLOBAL WARMING AT LEAST OCCASIONALLY

AND 33% OF AMERICAN ADULTS HEAR ABOUT GLOBAL WARMING IN THE MEDIA AT LEAST ONCE A WEEK

WHO WILL YOU TALK TO TODAY ABOUT CLIMATE CHANGE?

HOW TO DEBUNK CLIMATE CHANGE MISINFORMATION

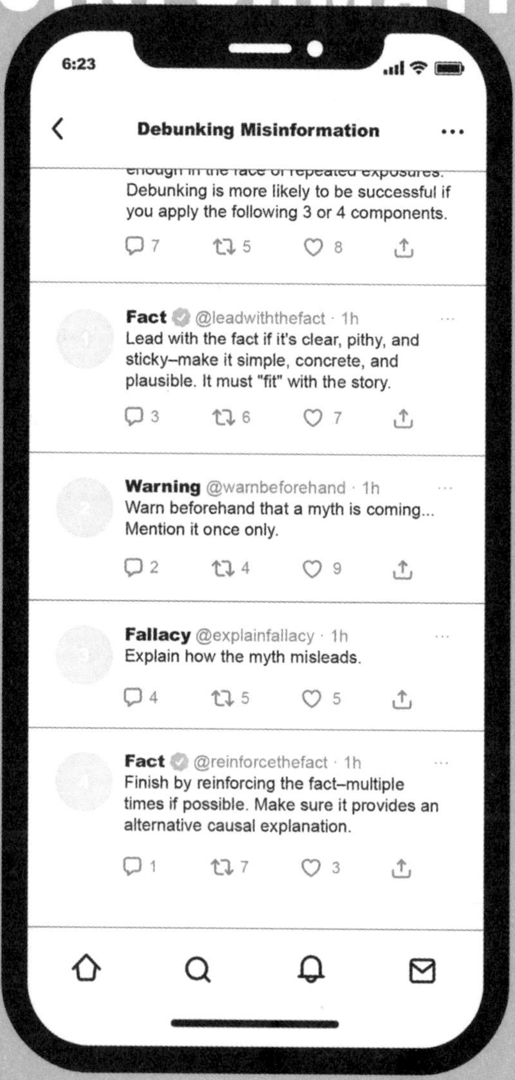

6:23
Debunking Misinformation
Debunking is more likely to be successful if you apply the following 3 or 4 components.
7 5 8
Fact @leadwiththefact · 1h
Lead with the fact if it's clear, pithy, and sticky–make it simple, concrete, and plausible. It must "fit" with the story.
3 6 7
Warning @warnbeforehand · 1h
Warn beforehand that a myth is coming... Mention it once only.
2 4 9
Fallacy @explainfallacy · 1h
Explain how the myth misleads.
4 5 5
Fact @reinforcethefact · 1h
Finish by reinforcing the fact–multiple times if possible. Make sure it provides an alternative causal explanation.
1 7 3

ACTION 98

GET SOCIAL ON SOCIAL MEDIA

Social media can be great but it's also beleaguered with a range of problems. When it comes to climate change, social media platforms can be a major source of misinformation and disinformation. A 2021 report by the environmental nonprofit Stop Funding Heat found that on Facebook alone there were up to 1.36 million views of climate misinformation each day! The report also found that only 3.6 percent of this misinformation was fact-checked by Facebook, which has promoted itself as having robust systems in place to counter misinformation and disinformation. Facebook (sometimes) directs users to its own Climate Science Center for fact-checked information when its algorithms detect mis- and disinformation on its feeds, but clearly it doesn't get rid of all misleading or downright inaccurate content.

While as individuals we can't control the methods Facebook, Twitter, or TikTok use to determine what content populates our feeds, we can point out false climate information and be careful about the information we share. The principle "see something, say something" can be especially powerful on social media. Most of us want accurate information, so when we call out something on social media as incorrect, it helps others become better informed. Of course, speaking up isn't always easy, but silence on these issues can yield the floor to an often vocal, misinformed minority.

Remember, part of what we see on our social media feeds depends on who we follow. If you are looking for some quality climate content, check out a few organizations and individuals who I think are worthy of a follow (and amplification) on Instagram, Twitter, and TikTok:

- @nasaclimatechange
- @unclimatechange
- @science_moms
- @futureearth
- @allwecansave
- @katharinehayhoe
- @ayanaeliza
- @icy_pete
- @noaaclimate
- @sunrisemvmt
- @ipcc_ch
- @natures

▶ **Will you say something when you spot climate misinformation or disinformation?**

ACTION 99

EMBRACE YOUR INNER BOOKWORM

Like most industries, the book publishing industry has an impact on global emissions, and many publishers, including the publisher of this book, are looking to decarbonize their supply chains and elevate their climate commitments. With over 825 million print books sold in the United States in 2021 and billions of books purchased globally each year, addressing the climate contributions from the book business—from production to shipping and delivery—can add up!

The energy source used for the creation of pulp and paper, in most cases, is the bulk of the emissions associated with books. In 2019, over 40 percent of Penguin Random House's carbon dioxide output came from the paper used in its books. This has many publishers looking to print more on recycled paper. But other aspects of the supply chain matter too, like the mode of transportation to get books from publisher to reader. Whether a train, a truck, or a ship is used for the bulk of a book's transportation from printer to consumer can shift a book's climate calculus. Penguin Random House (the parent company of this book's publisher), for example, works to reduce emissions across operations—from using renewable electricity in its warehouses and offices to aiming for more efficient shipping, including consolidation of orders and ensuring that trucks are full.

The next time you pick up a book, take a look at the publisher's name on the copyright page, then do some research online to see what climate-friendly practices the company has adopted. You can also check to see if the book was printed on recycled paper. The copyright page will also note whether the book has a Forest Stewardship Council (FSC) designation, which indicates that the paper the book is printed on comes from forests managed to meet a set of eco-friendly standards.

For bookworms, reading can also bring some climate benefits! Research suggests that reading climate fiction, a.k.a. cli-fi, can have "significant positive effects on readers' climate change beliefs and attitudes." If cli-fi isn't your preferred genre, new worthy climate nonfiction books are being published regularly too. Don't know where to start? Many universities, nonprofits, and advocacy groups have climate- and environment-related book clubs. The United Nations Sustainable Development Goals program even has a dedicated sustainability book club for children.

When you're ready to add some new titles to your reading list, here are a few to consider:

- *Drawdown: The Most Comprehensive Plan Ever Proposed to Reverse Global Warming* by Project Drawdown
- *Saving Us: A Climate Scientist's Case for Hope and Healing in a Divided World* by Katharine Hayhoe
- *All We Can Save: Truth, Courage, and Solutions for the Climate Crisis* edited by Ayana Elizabeth Johnson and Katharine K. Wilkinson
- *HERE: Poems for the Planet* edited by Elizabeth J. Coleman
- *The New Climate War: The Fight to Take Back Our Planet* by Michael Mann
- *No One Is Too Small to Make a Difference* by Greta Thunberg

▸ **What's next on your climate-focused reading list?**

ACTION 100

CELEBRATE SUCCESS AND EXPRESS GRATITUDE

Climate work—in all of its shapes, sizes, and scales—is important. We also know that more actions, by more people from all walks of life, are needed to confront the major climate change challenges ahead. If you've gotten this far in this book, you've probably already found quite a few climate actions to try, but this last action is one I hope we can all embrace and carry with us every day: Continue your climate solutions journey with an eye toward gratitude and celebration of success, no matter how big or small.

The realities of climate change can be daunting, emotional, and even terrifying. We may be experiencing loss, anxiety, or feeling totally overwhelmed. We may feel like every action doesn't actually matter. Many of these actions involve lots of research, sustained engagement and behavior change. In the face of these real challenges and barriers, I often return to my grandmother's wise words—"embrace an attitude of gratitude." Gratitude can help us wade through all of this messiness, while fueling and sustaining our own climate work and the work of others. A simple act of gratitude to yourself and to others can have a serious ripple effect.

While my grandmother wasn't a scientist, research supports her approach to life and challenging situations. Clinical psychology shows that acts of gratitude can actually foster and reinforce positive behaviors and actions. Recipients of gratitude are highly likely to perform the acknowledged behavior again in the future. So gratitude is not only an attitude, but also a way to multiply and sustain critical climate work.

Today, and on as many days as possible, engage in an act of gratitude. You can start small—an email, a letter, or a quick call is all it takes to thank someone for the climate work you admire. You can even write a note to yourself. If you want to scale that up, start a climate change thank-you-note-writing campaign. Bring together friends, colleagues, or community members to pepper people's inboxes or mailboxes with notes of thanks. Send them to local leaders, teachers, student groups, doctors, researchers, companies, farmers—everyone you see doing critical climate work. Yes, we need to demand more work be done, especially by our elected leaders, but how many folks get notes of thanks for the work they are currently doing? Want to go even bigger? Consider starting a local climate leadership award; giving credit to those leading the way can inspire others to join the cause.

As a fitting last act, I'd like to thank you for reading this book and for doing all you are doing on your own climate action journey. Whether you've taken one action or considered all 100, every action—including the act of reading this book—matters. The hope is that this handbook serves as a handy reminder of the broad diversity of work we can do and why it matters. As an individual, and as a collective, leaning into our strengths, pushing ourselves out of our comfort zones, and extending gratitude to those around us will help ensure we are making urgently needed progress towards a sustainable, healthy, and prosperous future for all. Thank you for your critical climate work.

▸ **Who will you thank for their climate work today?**

THANK YOU
THANK YOU
THANK YOU
THANK YOU
THANK YOU
THANK YOU
THANK YOU

CONTINUING YOUR CLIMATE JOURNEY

While some statistics presented in this book were perhaps surprising, others were likely sobering. I can attest that this book has changed how I show up to address climate change. It has informed how I can prioritize actions that work, right now, for my family and my community. We all have different strengths, capacities, and resources to engage in climate solutions. The diversity of ideas, information, and actions outlined here was intended to offer opportunities for each and every one of us to see ourselves as part of the solution—whether we are climate activists, highly climate-concerned, or just cautious about this whole climate change thing.

This is a thorny, complicated issue, but my hope is that we can all see ourselves in the solutions. We will certainly all benefit from the work we do today by knowing that we are making a brighter tomorrow at home, in our communities, and at scales bigger than ourselves.

We now need to find ways to revisit and sustain our climate work. Action and moving forward, even in small incremental steps, can help keep us motivated to keep up the work as we seek to create a better world around us. We know that we are close to locking in 1.5°C (2.7°F) of warming, and that there are costs and negative consequences associated with each additional amount of warming. At current rates of greenhouse gas emissions, most of us will see these targets exceeded within our lifetime. That means many of us stand to lose in the future from our choices today. Of course, the burden of these impacts is not equally distributed.

We all need to continue to listen, learn, and act. We need to lean into the challenge and continue to engage in diverse solutions. We need to act . . . again and again. Thank you once more for your commitment to climate solutions in all of their forms and across scales—at home, in your community, and beyond. Together, we can help secure a healthier, safer world for people and the planet. Our collective future depends on our climate actions today. Fortunately, the future isn't written in stone. We still choose our climate future, but the time to act is now.

▸ **What actions will you take?**

ACKNOWLEDGMENTS

This book would not be possible without the thousands of scientists, researchers, and students who have dedicated their lives to documenting, understanding, and developing solutions to climate change. Similarly, I'm thankful for the numerous individuals championing change in their communities, by employing their knowledge, passion, perspectives, and perseverance to build a better future for all of us. Thank you to my peers, colleagues, and mentors for your contributions, expertise, and commitment to this challenging work. Jen Worick, editor extraordinaire, helped to draw creativity from this climate scientist and dedicated an enormous amount of energy and expertise to this book. Thank you for being my partner in this adventure! While we created a book together, I'm beyond grateful for the friendship we've also developed through this process. Thanks to Joshua Powell, who used his incredible talents to bring this book to life through the power of art. I am immensely grateful to the whole team at Sasquatch Books for your willingness to see the forest through the trees with this project and for sharing your incredible expertise. Anna and Paul McKee, steadfast supporters and our Seattle family, helped to move this from idea into reality. To the team at the University of Minnesota Climate Adaptation Partnership: thank you for your support and patience through this project—I'm so honored to be working alongside you all to build climate resilience across landscapes and communities. Thank you to Connie and Peter Roop, my author role models and supportive parents who have never failed to be champions of my climate career, no matter where it took me in the world—even Antarctica! Last but not least, Peter, Abbie, and Kuna, this book would not be possible without your support, late night snack deliveries, and continued support as we work as a family to implement all 100 actions in this handbook. Words are insufficient to express my gratitude.

And to the readers of this book—thank you for your climate concern and, critically, your climate actions. Together, we can create a safe, prosperous, and thriving future for all. Onward toward action!

USEFUL CLIMATE TERMS

Common Terms

CARBON BUDGET: The total amount of carbon dioxide that can be emitted to stay below a certain global temperature.

CLIMATE CHANGE: The long-term change in Earth's average climate patterns including temperature, precipitation, wind, and other factors. Human-caused global warming is responsible for current observed climate changes.

(CLIMATE CHANGE) ADAPTATION: Actions that prepare or adjust to current and expected changes in climate. Climate change adaptation can occur across natural and human systems.

(CLIMATE CHANGE) MITIGATION: Actions or interventions that reduce the emission of greenhouse gases or enhance their removal from the atmosphere.

(CLIMATE CHANGE) RESILIENCE: The capacity or ability of human and natural systems to withstand or maintain function, respond to, and recover from climate change-related events and stressors.

GLOBAL WARMING: The long-term heating of Earth's surface due primarily to the increased concentration of heat-trapping greenhouse gases in Earth's atmosphere.

GLOBAL WARMING POTENTIAL: Describes the ability of a unit of gas to trap heat in the atmosphere over a set period of time. The global warming potential of different greenhouse gases is commonly calculated relative to carbon dioxide (CO_2) which is assigned a value of 1.

GREENHOUSE GAS: A heat-trapping gas present in Earth's atmosphere that is produced by natural and human activities. Common greenhouse gases in Earth's atmosphere include carbon dioxide (CO_2), methane (CH_4), nitrous oxide (N_2O), chlorofluorocarbons (CFC), and water vapor (H_2O). Each greenhouse gas remains in the atmosphere for different periods of time and has different abilities to absorb and reemit heat. These gases are produced by a range of different activities including the combustion of fossil fuels like coal, oil, and natural gas. Observed increases in atmospheric concentrations of greenhouse gases and the global warming they produce are predominately linked to human activities rather than natural causes, like volcanic eruptions.

NET-ZERO EMISSIONS: Achieved when the amount of human-produced greenhouse gases is matched by the amount being removed by activities that take greenhouse gases out of the atmosphere. Negative emissions occurs when the amount of greenhouse gases being removed exceeds the amount being produced.

OFFSETS: Credits, often tradable, for efforts that directly or indirectly reduce greenhouse gas emissions or compensate for avoided emissions of heat-trapping gases. Often referred to as carbon offsets or carbon credits.

PARIS AGREEMENT: A legally binding international treaty on climate change adopted by 196 countries in 2015, committing to the goal of limiting global temperature increase to well below 2°C (3.6°F) and to pursue efforts to limit global temperature increase to 1.5°C (2.7°F) above pre-industrial levels.

PRE-INDUSTRIAL: Can refer to any time period prior to the start of the Industrial Revolution. The most recent Intergovernmental Panel on Climate Change Assessment Reports use 1850-1900 as a pre-industrial baseline for global temperature and as a reference period against which global temperature change is compared. Different pre-industrial time periods are used in different studies and for different variables.

Abbreviations and Acronyms

BECCS bioenergy with carbon capture and storage
CCS carbon capture and storage
CH_4 methane
CO_2 carbon dioxide
DOE Department of Energy
DOI Department of the Interior
EIA Energy Information Administration
EPA Environmental Protection Agency
EV Electric Vehicle
FEMA Federal Emergency Management Agency
GHG Greenhouse Gas
GWP Global Warming Potential
IEA International Energy Agency
IPCC Intergovernmental Panel on Climate Change
NASA National Aeronautics and Space Administration
NOAA National Oceanic and Atmospheric Administration
N_2O Nitrous Oxide
PV Photovoltaics
UNFCCC United Nations Framework Convention on Climate Change
USGCRP United States Global Change Research Program

Common Units of Measure

CARBON DIOXIDE EQUIVALENT (CO_2e): A measure used to standardize the global warming potential (see terms list) of different greenhouse gases. This unit represents the number of metric tons of carbon dioxide emissions with the same global warming potential as one metric ton of a different greenhouse gas like methane.
KILOGRAM (kg): Equivalent to 2.2 pounds
METRIC TON (t): One thousand kilograms or 2204.62 pounds
MEGATON (Mt): A million metric tons
GIGATON (Gt): A billion metric tons

NOTES

PREFACE

"The Paris Agreement." United Nations Climate Change.

IPCC, 2018: Summary for Policymakers. In: Global Warming of 1.5°C. An IPCC Special Report on the impacts of global warming of 1.5°C above pre-industrial levels and related global greenhouse gas emission pathways, in the context of strengthening the global response to the threat of climate change, sustainable development, and efforts to eradicate poverty [Masson-Delmotte, V., P. Zhai, H.-O. Pörtner, D. Roberts, J. Skea, P.R. Shukla, A. Pirani, W. Moufouma-Okia, C. Péan, R. Pidcock, S. Connors, J.B.R. Matthews, Y. Chen, X. Zhou, M.I. Gomis, E. Lonnoy, T. Maycock, M. Tignor, and T. Waterfield (eds.)]. Cambridge University Press, Cambridge, UK and New York, NY, USA, pp. 3-24.

OVERVIEW

M. Allen et al., Warming caused by cumulative carbon emissions towards the trillionth tonne. *Nature* 458, 1163–1166, (2009).

Simon Evans, "Analysis: Which Countries Are Historically Responsible for Climate Change?," Carbon Brief, October 5, 2021.

IPCC, 2021: Climate Change 2021: The Physical Science Basis. Contribution of Working Group I to the Sixth Assessment Report of the Intergovernmental Panel on Climate Change[Masson-Delmotte, V., P. Zhai, A. Pirani, S.L. Connors, C. Péan, S. Berger, N. Caud, Y. Chen, L. Goldfarb, M.I. Gomis, M. Huang, K. Leitzell, E. Lonnoy, J.B.R. Matthews, T.K. Maycock, T. Waterfield, O. Yelekçi, R. Yu, and B. Zhou (eds.)]. Cambridge University Press, Cambridge, United Kingdom and New York, NY, USA, In press.

"GCP: Global Carbon Project: Homepage."

"Global CO_2 Emissions from Fossil Fuels and Land Use Change," Our World in Data.

"The Atmosphere: Getting a Handle on Carbon Dioxide—Global Climate Change: Vital Signs of the Planet."

Paul Griffin, "CDP Carbon Majors Report 2017," 2017.

IPCC, 2021: Summary for Policymakers. In: *Climate Change 2021: The Physical Science Basis. Contribution of Working Group I to the Sixth Assessment Report of the Intergovernmental Panel on Climate Change* [Masson-Delmotte, V., P. Zhai, A. Pirani, S.L. Connors, C. Péan, S. Berger, N. Caud, Y. Chen, L. Goldfarb, M.I. Gomis, M. Huang, K. Leitzell, E. Lonnoy, J.B.R. Matthews, T.K. Maycock, T. Waterfield, O. Yelekçi, R. Yu, and B. Zhou (eds.)]. Cambridge University Press, Cambridge, United Kingdom and New York, NY, USA, pp. 3–32.

US Environmental Protection Agency, 2021. Climate Change and Social Vulnerability in the United States: A Focus on Six Impacts. 430-R-21-003

"An Unfair Share: Exploring the Disproportionate Risks from Climate Change Facing Washington State Communities," Climate Impacts Group.

A. Leiserowitz et al., *Politics & Global Warming*, September 2021. Yale University and George Mason University. New Haven, CT: Yale Program on Climate Change Communication, (2018)

IPCC, 2018. Summary for Policymakers.

ACTION 1

Morten Fibieger Byskov, "Climate Change: Focusing on How Individuals Can Help Is Very Convenient for Corporations," *The Conversation*, January 10, 2019.

"Got Climate Doom? Here's What You Can Do to Actually Make a Difference," *New York Times*, November 10, 2021.

"The Powerful Role of Household Actions in Solving Climate Change," Project Drawdown, October 27, 2021.

ACTION 3

Attari Shahzeen Z. et al., "Public Perceptions of Energy Consumption and Savings," Proceedings of the National Academy of Sciences 107, no. 37: 16054–59, (2010).

Seth Wynes and Kimberly A. Nicholas, "The Climate Mitigation Gap: Education and Government Recommendations Miss the Most Effective Individual Actions," Environmental Research Letters 12, no. 7 (July 2017).

ACTION 4

"Fact Sheet: President Biden Sets 2030 Greenhouse Gas Pollution Reduction Target Aimed at Creating Good-Paying Union Jobs and Securing U.S. Leadership on Clean Energy Technologies," The White House, April 22, 2021.

"What Is the United Nations Framework Convention on Climate Change?," UNFCCC.

Simon Evans, "Analysis: Which Countries Are Historically Responsible for Climate Change?," Carbon Brief, October 5, 2021.

Cary Funk and Meg Hefferon, "U.S. Public Views on Climate and Energy," Pew Research Center Science (blog), November 25, 2019.

Leiserowitz et al., *Politics & Global Warming*, (2021).

ACTION 5.

IPCC, 2021. Summary for Policymakers.

"David Koch Was the Ultimate Climate Change Denier," *New York Times*, August 23, 2019.

"Exxon: The Road Not Taken," Inside Climate News (blog).

Naomi Oreskes and Erik M. Conway, Merchants of Doubt: How a Handful of Scientists Obscured the Truth on Issues from Tobacco Smoke to Global Warming (Bloomsbury Publishing USA, 2011).

Shannon Hall, "Exxon Knew about Climate Change Almost 40 Years Ago," *Scientific American*, October 26, 2015.

Mark Kaufman, "The Carbon Footprint Sham," Mashable India.

"Make Your Voice Heard! Stop the #AdsNotFit2Print!," Ads Not Fit To Print.

ACTION 6

"Buy Green Power," Minnesota Pollution Control Agency, November 16, 2009.

Hannah Ritchie, Max Roser, and Pablo Rosado, "Energy," Our World in Data, November 28, 2020.

ACTION 7

Ritchie, Roser, and Rosado, "Energy."

Max Roser, "Why Did Renewables Become so Cheap so Fast?," Our World in Data.

US Energy Information Administration, "In 2020, U.S. Coal Production Fell to Its Lowest Level since 1965" July 2021.

Chris Littlecott et al., "No New Coal by 2021: The Collapse of the Global Coal Pipeline," E3G, September 2021, 60.

"Levelized Cost of Energy and Levelized Cost of Storage 2019," Lazard, November 7, 2019.
"Solar, Wind and Battery Prices Falling," Climate Central, November 13, 2019.

ACTION 8

"Weather Power," Climate Central.
"Distributed Solar Photovoltaics," Project Drawdown.

ACTION 9

Kirsten Hund et al., "The Mineral Intensity of the Clean Energy Transition," 112, (2020).
Juzel Lloyd, "The Costs of Mining," *The Breakthrough Institute*, November 9, 2021.
Eric Lipton, Dionne Searcey, and Nicholas Forsythe, "What to Know About the Frantic Quest for Cobalt," *New York Times*, December 7, 2021.
Olivier Vidal, Bruno Goffé, and Nicholas Arndt, "Metals for a Low-Carbon Society," *Nature Geoscience* 6, no. 11 (November 1, 2013): 894–96.
Laura J. Sonter et al., 2020. "Renewable Energy Production Will Exacerbate Mining Threats to Biodiversity," Nature Communications 11, no. 1 (September 1, 2020): 4174.
Jack Healy and Mike Baker, "As Miners Chase Clean-Energy Minerals, Tribes Fear a Repeat of the Past," *New York Times*, December 27, 2021.
Jocelyn Timperley, "Explainer: These Six Metals Are Key to a Low-Carbon Future," Carbon Brief, April 12, 2018.
US Geological Survey "Mineral Commodity Summaries 2022.
Dionne Searcey et al., "A Power Struggle Over Cobalt Rattles the Clean Energy Revolution," *New York Times*, November 20, 2021.
Rachael Nealer, David Reichmuth, and Don Anair, "Cleaner Cars from Cradle to Grave: How Electric Cars Beat Gasoline Cars on LifetimeGlobal Warming Emissions," November 2015.

ACTION 10

N. McGuckin and A. Fucci, "Summary of Travel Trends: 2017 National Household Travel Survey," US Department of Transportation Federal Highway Administration, July 2018.
Wynes and Nicholas, "The Climate Mitigation Gap: Education and Government Recommendations Miss the Most Effective Individual Actions," 2017.
Henry Graber, "Millennials Are Driving Again, (but Not the Rich Ones,)" *Slate*, May 18, 2018.
Assuming use of a gas or petrol vehicle with an average fuel economy of 20-30 mpg; "Get Around Greener."
US Environmental Protection Agency, Greenhouse Gas Equivalencies Calculator.
United States Census Bureau,"Population Clock."
Somini Sengupta, "Trams, Cable Cars, Electric Ferries: How Cities Are Rethinking Transit," *New York Times*, October 3, 2021.

ACTION 11

Chad Frischmann and Crystal Chissell, "The Powerful Role of Household Actions in Solving Climate Change," Project Drawdown, October 27, 2021.
"Carpooling" Project Drawdown, February 6, 2020.
Jacob W. Ward, Jeremy J. Michalek, and Constantine Samaras, "Air Pollution, Greenhouse Gas, and Traffic Externality Benefits and Costs of Shifting Private Vehicle Travel to Ridesourcing Services," *Environmental Science & Technology* 55, no. 19 (October 5, 2021): 13174–85.

ACTION 12

US Environmental Protection Agency, "Fast Facts on Transportation Greenhouse Gas Emissions," 2020.
"The IEA Mobility Model—Programmes and Partnerships," International Energy Agency.
Brad Plumer, Nadja Popovich, and Blacki Migliozzi, "Electric Cars Are Coming. How Long Until They Rule the Road?" *New York Times*, March 10, 2021.
"Electric Vehicle Outlook 2021," BloombergNEF.
"Our Path to an All-Electric Future," General Motors.
Mike Scott, "Yes, Electric Cars Are Cleaner, Even When The Power Comes From Coal," *Forbes*, March 30, 2020.
Florian Knobloch et al., "Net Emission Reductions from Electric Cars and Heat Pumps in 59 World Regions over Time," *Nature Sustainability* 3, no. 6: 437–47, 7. (2021)
Hiroko Tabuchi and Brad Plumer, "How Green Are Electric Vehicles?" *New York Times*, March 2, 2021.
Drew Desilver, "Today's Electric Vehicle Market: Slow Growth in U.S., Faster in China, Europe," Pew Research Center.

ACTION 13

US Environmental Protection Agency, "Fast Facts on Transportation Greenhouse Gas Emissions."
US Department of Energy, "Driving More Efficiently."
"Drive Less and Drive Clean," Cool California.

ACTION 14

US Department of Energy, "Idling Reduction for Personal Vehicles," May 2015.
US Environmental Protection Agency, Greenhouse Gas Equivalencies Calculator.
US Environmental Protection Agency, "Idle-Free Schools Toolkit for a Healthy School Environment," Collections and Lists, July 1, 2019.
"Driving with the Start/Stop Function, Start/Stop Function, Starting and Driving, XC40 2021 Early, Volvo Support.

ACTION 15

Hannah Ritchie, "Climate Change and Flying: What Share of Global CO_2 Emissions Come from Aviation?" Our World in Data, October 22, 2020.
D.S. Lee et al., "The Contribution of Global Aviation to Anthropogenic Climate Forcing for 2000 to 2018," *Atmospheric Environment* 244 (January 1, 2021): 117834.
Stefan Gössling and Andreas Humpe, "The Global Scale, Distribution and Growth of Aviation: Implications for Climate Change," *Global Environmental Change* 65 (November 1, 2020): 102194.
Jocelyn Timperley, "Should We Give up Flying for the Sake of the Climate?"
Jamie Freed and Rajesh Kumar Singh, "Corporate business travel 'carbon budgets' loom for airlines," *Reuters*, October 11, 2021.

ACTION 16

World Tourism Organization and International Transport Forum (2019), *Transport-related CO_2 Emissions of the Tourism Sector—Modelling Results*, UNWTO, Madrid.
Manfred Lenzen et al., "The Carbon Footprint of Global Tourism," *Nature Climate Change* 8, no. 6: 522–28. (2018)
Brandon Graver, Kevin Zhang, and Dan Rutherford, "CO_2 Emissions from Commercial Aviation, 2018," 2019, 13.

US Environmental Protection Agency, 1.6 tCO_2e saved per roundtrip transatlantic flight; conversion from "Greenhouse Gas Equivalencies Calculator," 1.6 number from Wynes 2017.

ACTION 17

World Tourism Organization and United Nations Environment Programme, "Climate Change and Tourism—Responding to Global Challenges," 2008.
Danuta De Grosbois and David Fennell, "Carbon Footprint of the Global Hotel Companies: Comparison of Methodologies and Results," Tourism Recreation Research 36, no. 3 (January 1, 2011): 231–45.
US Environmental Protection Agency, assuming 175 kg/year, an average hotel room of 330 square feet. Values calculated using Greenhouse Gas Equivalencies Calculator.
Sustainable Hospitality Alliance, "Global Hotel Decarbonisation Report," November 2017.
Mingming Cheng et al., "The Sharing Economy and Sustainability—Assessing Airbnb's Direct, Indirect and Induced Carbon Footprint in Sydney," Journal of Sustainable Tourism 28, no. 8 (August 2, 2020): 1083-99

ACTION 18

D. Styles, H. Schönberger, and J.L. Galvez Martos, "Best Environmental Management Practice in the Tourism Sector," EUR 26022. Luxembourg (Luxembourg): Publications Office of the European Union; 2013. JRC82602.
"End of the Road for Bathroom Miniatures as IHG Opts for Bulk-Size Amenities to Reduce Plastic Waste," InterContinental Hotels Group PLC.
"Marriott International to Eliminate Single-Use Shower Toiletry Bottles From Properties Worldwide, Expanding Successful 2018 Initiative," August 28, 2019.
"California Bans Hotels from Using Tiny Plastic Bottles," *Associated Press*, October 9, 2019.

ACTION 19

"U.S. Travel Answer Sheet," U.S. Travel Association, March 2022.
PricewaterhouseCoopers, "Corporate Travel: Collaboration Is Essential for Successful COVID Recovery," PwC.
Kim M. Cobb, Peter Kalmus, and David M. Romps (2018), AGU should support its members who fly less, *Eos, 99*.
Judith T. Parrish (2017), Should AGU have fly-in meetings anymore?, *Eos, 98*.
US Environmental Protection Agency, Greenhouse Gas Equivalencies Calculator.
"U.S. Travel Answer Sheet," U.S. Travel Association, May 20, 2022.
Alana Semuels, "Business Travel's Demise Could Have Far-Reaching Consequences," *Time*, October 20, 2021.
Peter Caputo et al., "Return to a World Transformed: How the Pandemic Is Reshaping Corporate Travel," Deloitte Insights.
Jamie Freed and Rajesh Kumar Singh, "Analysis: Corporate Business Travel 'Carbon Budgets' Loom for Airlines," *Reuters*, October 11, 2021.

ACTION 20

Natalie Burg, "Who Funds the Fight Against Climate Change?," Means and Matters blog, April 20, 2022.
United Nations, "The Trillion Dollar Climate Finance Challenge (and Opportunity)," UN News, June 27, 2021.
"Financing Climate Action," United Nations.
Climate Policy Institute, "Global Landscape of Climate Finance 2021," December 2021.
Morgan Stanley, "Morgan Stanley Sustainable Signals: U.S. Individual Investors Maintain Strong Interest in Sustainable Investing Despite COVID-19 Pandemic."

ACTION 21

Dr Paul Griffin, "CDP Carbon Majors Report 2017," July 2017, 16.; In 2015, a fifth of global industrial GHG emissions was backed by publicly listed investment.
"Bank on our Future" pledge, Third Act.
"Banking on Climate Chaos 2022," Banking on Climate Chaos.
Andrew J. Hoffman and Todd Schifeling, "How Bill McKibben's Radical Idea of Fossil-Fuel Divestment Transformed the Climate Debate," *The Conversation*, December 11, 2017.
"Fossil Free: Divestment Frequently Asked Questions," Fossil Free: Divestment.
Atif Ansar, Ben Caldecott, and James Tilbury, "Stranded Assets and the Fossil Fuel Divestment Campaign: What Does Divestment Mean for the Valuation of Fossil Fuel Assets?," n.d., 81.
Noam Bergman, "Impacts of the Fossil Fuel Divestment Movement: Effects on Finance, Policy and Public Discourse," *Sustainability* 10, no. 7 (2018).
"Global Fossil Fuel Commitments Database."
David Carlin, "The Case For Fossil Fuel Divestment," *Forbes*, February 20, 2021.

ACTION 22

Jonathan I. Dingel and Brent Neiman, "How Many Jobs Can Be Done at Home?," National Bureau of Economic Research Working Paper Series, No. 26948, April 2020.
Kim Parker, Juliana Menasce Horowitz, and Rachel Minkin, "How the Coronavirus Outbreak Has—and Hasn't—Changed the Way Americans Work," Pew Research Center's Social & Demographic Trends Project (blog), December 9, 2020.
Alexander Michael Pearson, Tara Patel, and William Wilkes, 'Forever Changed': CEOs Are Dooming Business Travel—Maybe for Good," *Bloomberg*, August 31, 2021.
Dawit Habtemariam, "Survey: Virtual to Replace One-Quarter of 2022 Travel Volume, on Average," *Business Travel News*, July 26, 2021.
"Telepresence," Project Drawdown, February 6, 2020.
Carbon Trust Homeworking Report.
Sarah Holder, "The Environmental Implications of the Return to the Office," *Bloomberg*, March 29, 2021.
Rakesh Kochhar and Jeffrey S. Passel, "Telework May Save U.S. Jobs in COVID-19 Downturn, Especially among College Graduates," Pew Research Center (blog).

ACTION 23

Dr Sebastian Mueller et al., "Capturing the °Climate Factor | Linking Temperature Alignment and Financial Performance," Right. Based on Science (blog).
"Spark Ideas at Work," Breaking Boundaries.
"Climate Solutions at Work," Project Drawdown.

ACTION 24

"Embodied Carbon Call to Action Report," World Green Building Council.
Jessica Leung, "Decarbonizing U.S. Buildings," Center for Climate and Energy Solutions, July 2018.

US Energy Information Administration, "Frequently Asked Questions (FAQs)—US Energy Information Administration," February 2022.
Brian Clark Howard, "5 Surprising Ways Buildings Can Improve Our Health," *National Geographic*, February 14, 2017.
"Mayor de Blasio Signs Landmark Bill to Ban Combustion of Fossil Fuels in New Buildings," The official website of the City of New York, December 22, 2021.
OMS US Environmental Protection Agency, "High-Performance Buildings at EPA," Overviews and Factsheets, September 4, 2015.
LEED Rating System, US Green Building Council.

ACTION 25

Mariana Pickering, "6 Busted Green-Roof Myths," *Forbes*, December 12, 2013.
"Green and Cool Roofs," Project Drawdown.
US Department of Energy, "Benefits of Cool Roofs," Energy Saver blog.
"NYC CoolRoofs, NYC Business."
US Environmental Protection Agency, "Using Cool Roofs to Reduce Heat Islands," June 10, 2022.
Christina Poletto, "How 'Cool Roofs' Can Help Fight Climate Change," *New York Times*, November 10, 2021.
Leslie Guevarra, "DOE Installs 'Cool Roof' on DC Headquarters," GreenBiz, December 16, 2010.
"The Do-It-Yourself Ecoroof: A Do-It-Yourself Guide to Installing a Lightweight, Simple Sedum Ecoroof on a Shed-Sized Structure," City of Portland Environmental Services, 2014.
"Portland Ecoroofs," The City of Portland Environmental Services, 2022.

ACTION 26

Manish Ram et al., "Job Creation During a Climate Compliant Global Energy Transition across the Power, Heat, Transport, and Desalination Sectors by 2050," *Energy*, 238, Part A (January 2022): 121690. (2022)
Cara Buckley, "Coming Soon to This Coal County: Solar, in a Big Way," *New York Times*, January 2, 2022.
Sophie Yeo, "Clean Energy: The Challenge of Achieving a 'Just Transition' for Workers," CarbonBrief, January 4, 2017.
Jim Barrett and Josh Bivens, "The Stakes for Workers in How Policymakers Manage the Coming Shift to All-Electric Vehicles," Economic Policy Institute (EPI), September 22, 2021.
Savannah Bertrand, "Fact Sheet: Climate Jobs (2021)," Environment and Energy Study Institute (EESI), September 23, 2021.
Just Transition of the Workforce, and the Creation of Decent Work and Quality Jobs, United Nations Framework Convention on Climate Change, October 26, 2016.
"Johns Hopkins University and American Flood Coalition New Economic Study Reveals $1B in Flood Resilience Investment Could Create up to 40,000 New Jobs," American Flood Coalition, December 8, 2020.
Jenessa Duncombe, "Climate Change Will Make Us Sicker and Lose Work Hours," *EOS*, November 18, 2019.
Lancet Countdown, 2019: 2019 Lancet Countdown on Health and Climate Change Policy Brief for the United States of America. Salas RN, Knappenberger P, Hess JJ. Lancet Countdown U.S. Policy Brief, London, United Kingdom, p. 11.
"Chapter 21: Midwest," Fourth National Climate Assessment, 2018.
"Key Findings," The New Climate Economy, The 2018 Report of the Global Commission on the Economy and Climate.

ACTION 27

"More Than 1,000 Companies Commit to Science-Based Emissions Reductions in Line with 1.5°C Climate Ambition," Science Based Targets, November 10, 2021.
Net Zero Tracker, ZeroTracker.net.
IPCC, 2018. "Index" in *Global Warming of 1.5°C. An IPCC Special Report on the Impacts of Global Warming of 1.5°C above Pre-Industrial Levels and Related Global Greenhouse Gas Emission Pathways, in the Context of Strengthening the Global Response to the Threat of Climate Change, Sustainable Development, and Efforts to Eradicate Poverty.*
"The Paris Agreement," United Nations Climate Change.
Taylor Francis, "Companies Need to Have Climate Plans—Not Just Climate Pledges," *Fast Company*, December 16, 2021.
"Why Corporate Climate Pledges of 'Net-Zero' Emissions Should Trigger a Healthy Dose of Skepticism," *The Conversation*, March 25, 2021.
Ceri Perkins, "Are Companies' Climate Pledges for Real—Or Just Hot Air? Here's How You Can Tell," Ideas.TED.com, June 16, 2021.
Michael Corkery and Julie Creswell, "Corporate Climate Pledges Often Ignore a Key Component: Supply Chains," *New York Times*, November 3, 2021.

ACTION 28

IPCC, 2018. Special Report: Global Warming of 1.5°C
IPCC, 2019. Summary for Policymakers in Climate Change and Land: An IPCC Special Report on Climate Change, Desertification, Land Degradation, Sustainable Land Management, Food Security, and Greenhouse Gas Fluxes in Terrestrial Ecosystems, 2019. [P.R. Shukla, J. Skea, E. Calvo Buendia, V. Masson-Delmotte, H.- O. Pörtner, D. C. Roberts, P. Zhai, R. Slade, S. Connors, R. van Diemen, M. Ferrat, E. Haughey, S. Luz, S. Neogi, M. Pathak, J. Petzold, J. Portugal Pereira, P. Vyas, E. Huntley, K. Kissick, M. Belkacemi, J. Malley, (eds.)] In press.
US Environmental Protection Agency, Greenhouse Gas Equivalencies Calculator.

ACTION 29

"Food Waste: What's at Stake," National Resources Defence Council.
Jean C. Buzby, Hodan Farah Wells, and Jaspreet Aulakh, "Food Loss—Questions about the Amount and Causes Still Remain," June 2, 2014, US Department of Agriculture, Economic Research Service.
IPCC, 2019. Summary for Policymakers.
"Global Issues: Food," United Nations.

ACTION 30

Dana Gunders, Wasted: How America Is Losing up to 40 Percent of Its Food from Farm to Fork to Landfill, National Resources Defense Council, August 2012.
IPCC, 2021. Summary for Policymakers. In: Climate Change 2021: The Physical Science Basis.
US Environmental Protection Agency, Composting at Home, last updated July 7, 2022.
US Environmental Protection Agency, Reducing the Impact of Wasted Food by Feeding the Soil and Composting, last updated February 12, 2022.

ACTION 31

Brent R. Heard et al., "Comparison of Life Cycle Environmental Impacts from Meal Kits and Grocery Store Meals," *Resources, Conservation and Recycling* 147 (August 2019): 189–200.

Springer Nature.

A. Goodchild, A., E. Wygonik, E. and N. Mayes, An analytical model for vehicle miles traveled and carbon emissions for goods delivery scenarios. Eur. Transp. Res. Rev. 10, 8 (2018).

ACTION 32

Levi Sumagaysay, "The Pandemic Has More Than Doubled Food-Delivery Apps' Business. Now What?" MarketWatch, updated November 27, 2020.

S. Lock, "Amount of Smartphone Food Delivery App Users in the U.S. 2019–2023," *Statista*, September 6, 2021.

Robert Crawford, "Home-Delivered Food Has a Huge Climate Cost. So which cuisine is the worst culprit?" *The Conversation*, January 18, 2021.

Sally Ho, "4 Surprising Things to Know about the Carbon Footprint of Your Online Food Delivery," Green Queen, March 11, 2021.

Indumathi Arunan and Robert H. Crawford, "Greenhouse Gas Emissions Associated with Food Packaging for Online Food Delivery Services in Australia," *Resources, Conservation and Recycling* 168 (May 2021): 105299.

Alejandro Gallego-Schmid, Joan Manuel F. Mendoza, and Adisa Azapagic, "Environmental Impacts of Takeaway Food Containers," *Journal of Cleaner Production* 211 (February 20, 2019): 417–27.

Jybe App, Jybe, Inc.

ACTION 33

J. Poore and T. Nemecek, "Reducing Food's Environmental Impacts through Producers and Consumers," *Science* 360, no. 6392 (June 1, 2018): 987–92.

Saloni Shah and Dan Blaustein-Rejto, "Federal Support for Alternative Protein for Economic Recovery and Climate Mitigation," *Breakthrough Institute*, May 12, 2020.

Saloni Shah and Dan Blaustein-Rejto, "Can Cultivated Meat Live up to Its Environmental Promise?," *Breakthrough Institute,* May 13, 2020.

"About Us," *Breakthrough Institute.*

Linus Blomqvist, "Eat Meat. Not Too Much. Mostly Monogastrics," *Breakthrough Institute*, January 29, 2019.

ACTION 34

Meagan Bridges, "Moo-ve Over, Cow's Milk: The Rise of Plant-Based Dairy Alternatives," *Practical Gastroenterology* 171 (January 2018).

Nathan Clay, Alexandra E. Sexton, Tara Garnett, and Jamie Lorimer, "Palatable Disruption: The Politics of Plant Milk," *Agriculture and Human Values* 37, no. 4 (January 30, 2020): 945–962.

Swati Sethi, S. K. Tyagi, and Rahul K. Anurag, "Plant-Based Milk Alternatives an Emerging Segment of Functional Beverages: A Review," *Journal of Food Science and Technology* 53, no. 9 (September 2, 2016): 3408–3423.

D. L. Liebe, M. B. Hall, and R. R. White, "Contributions of Dairy Products to Environmental Impacts and Nutritional Supplies from United States Agriculture," *Journal of Dairy Science* 103, no. 11 (November 2020): 10867–10881.

J. Poore and T. Nemecek, "Reducing Food's Environmental Impacts through Producers and Consumers," *Science* 360, no. 6392 (June 1, 2018): 987–92.

Annette McGivney, "Almonds Are Out. Dairy Is a Disaster. So What Milk Should We Drink?," *The Guardian*, January 29, 2020.

Annelies Boerema et al., 2016. "Soybean Trade: Balancing Environmental and Socio-Economic Impacts of an Intercontinental Market," *PloS ONE* 11, no. 5 (May 31, 2016).

Robin R. White. and Mary Beth Hall, "Nutritional and Greenhouse Gas Impacts of Removing Animals from US Agriculture," *Proceedings of the National Academy of Sciences* 114, no. 48 (November 13, 2017).

ACTION 35

Gregory R. Peterson, "Is Eating Locally a Moral Obligation?," *Journal of Agricultural and Environmental Ethics* 26 (2013): 421–437.

Carola Grebitus, Jayson L. Lusk, and Rodolfo M. Nayga, "Effect of Distance of Transportation on Willingness to Pay for Food," *Ecological Economics* (April 1, 2013): 67–75.

Sonja Brodt et al., "Comparing Environmental Impacts of Regional and National-Scale Food Supply Chains: A Case Study of Processed Tomatoes," *Food Policy* 42 (October 1, 2013): 106–14, (2013).

Libby O. Christensen, Ryan E. Galt, and Alissa Kendall, "Life-Cycle Greenhouse Gas Assessment of Community Supported Agriculture in California's Central Valley," *Renewable Agriculture and Food Systems* 33, no. 5 (2017): 393–405.

ACTION 36

Statisia Research Department, "U.S. Population: Consumption of Chocolate and Other Candy from 2011 to 2024," Statisia, November 2020.

Niall McCarthy, "The World's Biggest Chocolate Consumers," *Forbes*, July 22, 2015.

Michon Scott, "Climate & Chocolate," Climate.gov, February 10, 2016.

Dave Reay, "Climate-Smart Chocolate," in Climate-Smart Food, ed. Dave Reay (Cham: Springer International Publishing, 2019), 69–79.

Lauren Okinaka, "5 Must-See Cacao Farms on Hawai'i Island," *Hawai'i Magazine*, July 7, 2021.

Sierra Garcia, "Will Chocolate Survive Climate Change? Actually, Maybe," JSTOR Daily, August 31, 2021.

ACTION 37

Climate Matters, "Climate Change & Wine," Climate Central, September 29, 2021.

Matt Strickland, "Climate Change and Spirits: The Effects on Alcohol Agriculture," *Distiller*, October 10, 2019.

M. A White et al., "Extreme Heat Reduces and Shifts United States Premium Wine Production in the 21st Century," *Proceedings of the National Academy of Sciences* 103, no. 30 (July 25, 2006): 11217–22.

Lee Hannah et al., "Climate Change, Wine, and Conservation," *Proceedings of the National Academy of Sciences* 110, no. 17 (April 23, 2013): 6907–12.

"Will Your Favorite Drink Survive Climate Change?" Zurich, September 16, 2020.

Climate: Our Commitments, New Belgium Brewing.

Better World Shopper.

Brewers' Climate Declaration, Business Climate Leaders.

Wine Industry's Climate Declaration, Business Climate Leaders.

Carbon Calculator for Breweries, Appalachian Offsets.

ACTION 38

Carmen Nab and Mark Maslin, "Life Cycle Assessment Synthesis of the Carbon Footprint of Arabica Coffee: Case Study of Brazil and Vietnam Conventional and Sustainable Coffee Production and Export to the United Kingdom," *Geo: Geography and Environment* 7, no. 2 (July–December 2020).

Gino B. Bianco, "Climate Change Adaptation, Coffee, and Corporate Social Responsibility: Challenges and Opportunities," *International Journal of Corporate Social Responsibility* 5, no. 3 (August 2020).

Royal Geographical Society.

B. Killian, L. Rivera, M. Soto., and D. Navichoc, (2013). Carbon Footprint across the Coffee Supply Chain: The Case of Costa Rican Coffee. *Journal of Agricultural Science and Technology,* 3, 151–170.

ACTION 39

IPCC, 2019. Special Report on the Ocean and Cryosphere in a Changing Climate.

António Marques et al., "Climate Change and Seafood Safety: Human Health Implications," *Food Research International* 43, no. 7 (August 1, 2010): 1766–79.

Michael J. MacLeod et al., "Quantifying Greenhouse Gas Emissions from Global Aquaculture," *Scientific Reports* 10, (July 15, 2020): 11679.

US Environmental Protection Agency, Greenhouse Gas Equivalencies Calculator.

Consider Climate, Monterey Bay Aquarium Seafood Watch.

J. Poore and T. Nemecek, "Reducing Food's Environmental Impacts through Producers and Consumers," *Science* 360, no. 6392 (June 1, 2018): 987–92.

Elizabeth M.P. Madin and Peter I. Macreadie, "Incorporating Carbon Footprints into Seafood Sustainability Certification and Eco-Labels," *Marine Policy* 57 (July 1, 2015): 178–81.

ACTION 40

IRP, 2018. *Re-defining Value—The Manufacturing Revolution. Remanufacturing, Refurbishment, Repair and Direct Reuse in the Circular Economy.* Nabil Nasr, Jennifer Russell, Stefan Bringezu, Stefanie Hellweg, Brian Hilton, Cory Kreiss, and Nadia von Gries. A Report of the International Resource Panel. United Nations Environment Programme, Nairobi, Kenya.

ACTION 41

K. Niinimäki et al., The environmental price of fast fashion. Nature Reviews Earth Environment 1, 189–20, (2020).

Andrew Morlet et al., "A New Textiles Economy: Redesigning Fashions' Future" Ellen MacArthur Foundation, updated January 12, 2017.

Quantis 2018.

Jarkko Levänen et. al, "Innovative recycling or extended use? Comparing the global warming potential of different ownership and end-of-life scenarios for textiles," Environmental Research Letters 16 054069, 2021.

ACTION 42

Wynes and Nicholas, "The Climate Mitigation Gap: Education and Government Recommendations Miss the Most Effective Individual Actions."

US Environmental Protection Agency, Greenhouse Gas Equivalencies Calculator.

Kendyl Salcito, "Why Cloth Diapers Might Not Be the Greener Choice, After All," *Washington Post,* May 8, 2015.

US Environmental Protection Agency, "Nondurable Goods: Product-Specific Data," last updated December 14, 2021.

UK Department for Environment, Food and Rural Affairs, Updated Life Cycle Study on Reusable and Disposable Nappies.

ACTION 43

Robbe Geerts et al., "Bottle or Tap? Toward an Integrated Approach to Water Type Consumption," *Water Research* 173, no. 15 (April 2020): 115578.

Peter H. Gleick, Bottled and Sold: The Story Behind Our Obsession with Bottled Water (Washington DC: Island Press, 2010).

Zhihua Hu, Lois Wright Morton, and Robert L. Mahler, "Bottled Water: United States Consumers and Their Perceptions of Water Quality," *International Journal of Environmental Research and Public Health* 8, no. 2 (February 2011): 565–578.

Barbara Harfmann, "2021 State of the Beverage Industry: Bottled Water Bubbling Over with Growth," Beverage Industry, July 1, 2021.

Eastern Research Group, Life Cycle Assessment of Drinking Water Systems: Bottle Water, Tap Water, and Home/Office Delivery Water, State of Oregon Department of Environmental Quality, October 22, 2009.

Valentina Fantin et al., "A Method for Improving Reliability and Relevance of LCA Reviews: The Case of Life-Cycle Greenhouse Gas Emissions of Tap and Bottled Water," *Science of the Total Environment* 476–477 (April 2014): 228–241.

Geerts, 2020. "Bottle or Tap? Toward an Integrated Approach to Water Type Consumption," and references therein.

ACTION 44

US Energy Information Administration, FAQs: How Much Oil Is Used to Make Plastic?, last reviewed June 1, 2021.

Lisa Anne Hamilton and Steven Feit et al., Plastic & Climate: The Hidden Costs of a Plastic Planet, Center for International Environmental Law (CIEL), May 2019.

Annual Emissions from the Plastic Lifestyle infographic, Center for International Environmental Law (CIEL), 2019.

From an interview with Kerri Major, "Plastic Waste and Climate Change—What's the Connection?" World Wildlife Fund Australia, June 30, 2021.

Brooke Bauman, "How Plastics Contribute to Climate Change," Yale Climate Connections, August 20, 2019.

ACTION 45

National Oceanic and Atmospheric Administration, "What Are Microplastics?" last updated February 26, 2021.

A. Kelly et al., "Microplastic Contamination in East Antarctic Sea Ice," *Marine Pollution Bulletin* 154 (May 2020): 111130.

K. Kvale et al., "Zooplankton Grazing of Microplastic Can Accelerate Loss of Ocean Oxygen," *Nature Communications* 12 (April 2021): 2358.

Helmholtz Centre for Ocean Research Kiel (GEOMAR), "Microplastics Affect Global Nutrient Cycle and Oxygen Levels in the Ocean," April 21, 2021.

J. Ming and F. Wang, "Microplastics' Hidden Contribution to Snow Melting," *Eos,* March 8, 2021.

ACTION 46

Miguel Jaller and Anmol Pahwa, "Evaluating the Environmental Impacts of Online Shopping: A Behavioral and Transportation Approach," *Transportation Research Part D: Transport and Environment* 80, Science Direct, (March 2020).

Amy Haimerl, "When You're a Small Business, E-Commerce Is Tougher Than It Looks," *New York Times*, updated August 24, 2021.

Adele Peters, "Can Online Retail Solve Its Packaging Problem?" *Fast Company*, April 20, 2018.

Hanne Siikavirta et al., "Effects of E-Commerce on Greenhouse Gas Emissions: A Case Study of Grocery Home Delivery in Finland," *Journal of Industrial Ecology* 6, no. 2 (April 2002); 83–97.

Miguel Jaller, "How E-Commerce Is Reshaping Warehousing and Impacting Disadvantaged Communities—And What We Can Do about It," UC Davis Institute of Transportation Studies, June 29, 2020.

Graham Charlton, "Ecommerce Returns: 2020 Stats and Trends," SaleCycle, January 15, 2020, last revision by Brad Ward July 8, 2021.

Miguel Jaller, Anmol Pahwa, and Seth Karten, "UC Davis Transportation and Climate Blog: Keeping e-Commerce Environmentally Friendly—What Consumers Can Do," UC Davis Institute of Transportation Studies Sustainable Freight, December 2, 2020.

ACTION 47

Miguel Jaller and Anmol Pahwa, "Analytical Modeling Framework to Assess the Economic and Environmental Impacts of Residential Deliveries, and Evaluate Sustainable Last-Mile Strategies," UC Davis: National Center for Sustainable Transportation Research Report (March 2020).

Jaller, Pahwa, and Karten, "UC Davis Transportation and Climate Blog: Keeping e-Commerce Environmentally Friendly—What Consumers Can Do," December 2, 2020.

Jaller and Pahwa, "Evaluating the Environmental Impacts of Online Shopping: A Behavioral and Transportation Approach," March 2020.

ACTION 48

Lotfi Belkhir, "How Smartphones Are Heating Up the Planet," *The Conversation*, March 25, 2018.

Lotfi Belkhir and Ahmed Elmeligi, "Assessing ICT Global Emissions Footprint: Trends to 2040 & Recommendations," *Journal of Cleaner Production* 177 (March 2018): 448–463.

Sandra Wilson, "I'm a Bit of a Modern-Day Alchemist, Recovering Gold from Old Mobile Phones," *The Conversation*, May 21, 2020.

Josh Lepawsky, "Almost Everything You Know about E-Waste is Wrong," *The Conversation*, May 15, 2018.

Ian Williams, "Our Thirst for New Gadgets Has Created a Vast Empire of Electronic Waste," *The Conversation*, February 10, 2016

Sunil Heart, "E-Waste: The High Cost of High-Tech," *The Conversation*, December 11, 2011.

E-Waste Monitors, United Nations Institute for Training and Research (UNITAR).

Ainissa Ramirez, "Where to Find Rare Earth Elements," NOVA, April 2, 2013.

ACTION 49

Kenneth Carling et al., "Measuring Transport-Related CO_2 Emissions Induced by Online and Brick-and-Mortar Retailing," *Transportation Research Part D: Transport and Environment* 40: 28–42, 2015.

Ellis Jones, "Can We Quantify Ethical Consumption? Reflections on a Decade-Long Exercise in Public Sociology," *Humanity & Society* 45, no 1.2021.

ACTION 50

"The Sustainability Imperative," NielsenIQ, October 12, 2015.

Lydia Noyes, "A Guide to Greenwashing and How to Spot It," *EcoWatch*, October 21, 2021

Merriam-Webster, s.v. "greenwashing (n)."

Ambitious Corporate Climate Action, Science Based Targets.

Net Zero Tracker, ZeroTracker.net.

Tom Lyon and Magali (Maggie) Delmas, "When Corporations Take Credit for Green Deeds, Their Lobbying May Tell Another Story," *The Conversation*, updated April 23, 2019.

Environmentally Friendly Products: FTC's Green Guides, Federal Trade Commission.

ACTION 51

"Record Number of Billion-Dollar Disasters Struck U.S. in 2020," National Oceanic and Atmospheric Administration (NOAA), January 8, 2021.

Risk Rating 2.0: Equity in Action, Federal Emergency Management Agency (FEMA), last updated April 18, 2022.

Christopher Flavelle, "The Cost of Insuring Expensive Waterfront Homes Is About to Skyrocket," *New York Times*, September 24, 2021.

"Report Chapter List," Fourth National Climate Assessment, 2018.

ACTION 52

Climate Change and Your Insurance, Office of the Insurance Commissioner Washington State.

United Nations Principles for Sustainable Insurance (PSI) and US National Association of Insurance Commissioners (NAIC), "COP26 PSI—How Are US Insurance Regulators Taking Action on Climate Change Risks? Webinar," BrightTalk, November 3, 2021.

How Climate Change Affects Your Insurance, Office of the Insurance Commissioner Washington State.

"Global Insured Catastrophe Losses Rise to USD 112 Billion in 2021, the Fourth Highest on Record, Swiss Re Estimates," Swiss Re Group press release, December 14, 2021.

"California Issues 1-Year Moratorium on Homeowner Insurance Cancellations and Non-Renewals," Insurance Journal, September 20, 2021.

"What You Should Know about Named Storm Deductibles," National Association of Insurance Commissioners (NAIC), July 12, 2021.

Federal Emergency Management Agency, "FEMA Updates Its Flood Insurance Rating Methodology to Deliver More Equitable Pricing," news release no. HQ-21-079, April 1, 2021.

Risk Rating 2.0: Projected Premium Changes by Zip Code—SFH Policies, ARCgis online.

Laura Lightbody, "New Interactive Map Shows Impacts of Federal Flood Insurance Rate Changes," PEW, September 20, 2021.

ACTION 53

International Federation of Red Cross (IFRC), World Disasters Report 2020: Come Heat or High Water—Tackling the Humanitarian Impacts of the Climate Crisis Together," ReliefWeb, November 17, 2020.
Build a Kit, Ready.gov, last updated May 5, 2022.
A.C. Shilton, "Pack a 'Go Bag' Now," *New York Times,* September 19, 2020.

ACTION 54

US Environmental Protection Agency, "Health, Energy Efficiency and Climate Change," last updated October 5, 2021.
Benjamin Goldstein, Dimitrios Gounaridis, and Joshua P. Newell, "The Carbon Footprint of Household Energy Use in the United States," Proceedings of the National Academy of Sciences (PNAS), July 20, 2020.

ACTION 55

"Clean Cooking Can Reduce Pollution from Burning Wood or Coal in Traditional Stoves and Protect Human Health," Project Drawdown.
US Energy Information Administration, Residential Energy Consumption Survey (Recs), revised May 2018.
Brian Clark Howard, "5 Surprising Ways Buildings Can Improve Our Health," *National Geographic,* February 14, 2017.
Sabrina Imbler, "Kill Your Gas Stove," *The Atlantic,* October 15, 2020.
US Environmental Protection Agency, Basic Information about NO2, last updated August 2, 2022.
Brady Seals, "Indoor Air Pollution: The Link Between Climate and Health," RMI, May 5, 2020.
Paul Hope, "Pros and Cons of Induction Cooktops and Ranges," *Consumer Reports,* December 3, 2019.
"High-Efficiency Heat Pumps," Project Drawdown.
US Environmental Protection Agency, "Study Finds Exposure to Air Pollution Higher for People of Color Regardless of Region or Income," September 20, 2021.

ACTION 56

Chiara Delmastro, Cooling, International Energy Agency (IEA), November 2021.
Dr. Fatih Birol, "The World's Electricity Systems Must Be Ready to Counter the Growing Climate Threat," International Energy Agency (IEA), July 12, 2021.
The Future of Cooling, International Energy Agency (IEA), May 2018.
"Climate-Friendly Cooling Could Cut Years of Greenhouse Gas Emissions and Save Trillions of Dollars, International Energy Agency (IEA), July 17, 2020.
"Home Energy Use," Center for Climate and Energy Solutions (C2ES).

ACTION 57

"Renewable Electricity Growth Is Accelerating Faster Than Ever Worldwide, Supporting the Emergence of the New Global Energy Economy," International Energy Agency (IEA), December 1, 2021.
United Nations Climate Change, Unfccc.int.
Yasmina Abdelilah et al., Renewables 2021: Analysis and Forecast to 2026, International Energy Agency (IEA), December 2021.
Simon Evans, "Solar Is Now 'Cheapest Electricity in History,' Confirms IEA," CarbonBrief, October 13, 2020.
US Environmental Protection Agency, "Distributed Generation of Electricity and its Environmental Impacts," July 14, 2021
Homeowner's Guide to the Federal Tax Credit for Solar Photovoltaics, US Department of Energy, Office of Energy Efficiency & Renewable Energy
"Distributed Solar Photovoltaics," Project Drawdown.
US Department of Energy, "Heat Pump Water Heaters," Energy Saver blog.
"Solar Hot Water," Project Drawdown.

ACTION 58

US Department of Energy, "LED Lighting," Energy Saver blog.
"LED Lighting," Project Drawdown.
US Department of Energy, "Lighting Choices to Save You Money," Energy Saver blog.
John Schwartz, "White House to Relax Energy Efficiency Rules for Light Bulbs," *New York Times,* updated September 6, 2019.
US Environmental Protection Agency, Carbon Footprint Calculator, last updated July 14, 2016.
US Environmental Protection Agency, Greenhouse Gas Equivalencies Calculator.

ACTION 59

Chad Frischmann and Crystal Chissell, "The Powerful Role of Household Actions in Solving Climate Change," Project Drawdown, October 27, 2021.
IPCC, Special Report: Global Warming of 1.5°C.
Amy Snover et al., "No Time to Waste: The IPCC Special Report on Global Warming of 1.5°C and Implications for Washington State," University of Washington Climate Impacts Group, updated February 2019.
"Low-Flow Fixtures," Project Drawdown.
Diana M. Ruiz et al., "Turning Off the Tap: Common Domestic Water Conservation Actions Insufficient to Alleviate Drought in the United States of America," *PloS One* 15, no. 3 (March 2020): e0229798.

ACTION 60

"Home Energy Use," Center for Climate and Energy Solutions (C2ES).
US Energy Department, Energy Efficiency Program: Energy Conservation Standards for Residential Clothes Washers proposed rule, Federal Register, August 2, 2019.
Joanna Mauer, "A New Spin on Clothes Washer Efficiency Coming in January 2018," Appliance Standards Awareness Project (ASAP), December 27, 2017.
Clothes Washers, Energy Star.
Wynes and Nicholas, "The Climate Mitigation Gap: Education and Government Recommendations Miss the Most Effective Individual Actions."
USDA Plant Hardiness Zone Map, US Department of Agriculture Agricultural Research Service.

ACTION 61

"Home Energy Use," Center for Climate and Energy Solutions (C2ES).
Climate Matters, "Planting Zones Moving North," Climate Central, April 10, 2019.
Christopher W. Woodall, Research Review No. 11, US Forest Service Northern Research Station, Autumn 2010.
Nature's Notebook, US Geological Survey and USA National Phenology Network.
"Leaf Blower's Emissions Dirtier Than High-Performance Pick-Up Truck's, Says Edmunds' InsideLine.com," Edmunds, December 6, 2011.
Margaret Renkl, "The First Thing We Do, Let's Kill All the Leaf Blowers," *New York Times,* October 25, 2021.

Clarasteyer, "Electric or Gas Leaf Blowers . . . Neither?," Washington University in St. Louis, November 17, 2020.
Phil Willon, "California Moves Toward Ban on Gas Lawn Mowers and Leaf Blowers," *Los Angeles Times*, October 9, 2021.

ACTION 62

US Environmental Protection Agency, National Overview: Facts and Figures on Materials, Wastes and Recycling, last updated July 31, 2021.
Laura Sullivan, "How Big Oil Misled the Public into Believing Plastic Would Be Recycled," *NPR*, September 11, 2020.
US Environmental Protection Agency, Plastics: Material-Specific Data, last updated September 30, 2021.
"Recycling," Project Drawdown.
Adele Peters, "All the Ways Recycling Is Broken and How to Fix Them," *Fast Company*, April 4, 2019.
Mark C. Crowley, "Remote Work Has a Downside. Here's Why I Want to Go Back to the Office," *Fast Company*, November 30, 2021.

ACTION 63

US Environmental Protection Agency, Household Carbon Footprint Calculator.

ACTION 64

IPCC, 2018.
National Academies of Sciences, Engineering, and Medicine, A Research Strategy for Ocean-Based Carbon Dioxide Removal and Sequestration (Washington, DC: The National Academies Press, 2022), 39–76.
Amy Snover et al., "No Time to Waste: The IPCC Special Report on Global Warming of 1.5°C and Implications for Washington State."
Christoph Beuttler, Louise Charles, and Jan Wurzbacher, "The Role of Direct Air Capture in Mitigation of Anthropogenic Greenhouse Gas Emissions," Frontiers in Climate, November 21, 2019.
Professor Kirsten Zickfeld, "Guest Post: Why CO_2 Removal Is Not Equal and Opposite to Reducing Emissions," CarbonBrief, June 21, 2021.
Sarah Budinis, Direct Air Capture, International Energy Agency (IEA), November 2021.
Oliver Mitenberger and Matthew D. Potts, "Why Corporate Climate Pledges of 'Net-Zero' Emissions Should Trigger a Healthy Dose of Skepticism," *The Conversation*, March 25, 2021.
Brian Palmer, "Should You Buy Carbon Offsets?," National Resources Defense Council (NRDC), May 11, 2022.

ACTION 65

US Department of Energy, Office of Energy Efficiency & Renewable Energy, "Bioenergy Basics."
Jocelyn Timperley, "In-Depth: The Challenge of Using Biofuels to Cut Transport Emissions," CarbonBrief, July 19, 2017.
US Energy Information Administration, "Biofuels Explained," last updated April 7, 2022.
Emily Pontecorvo, "Biofuels Are a Controversial Climate Solution. Could They Still Help Save the Planet?," Grist, September 11, 2020.
IPCC, 2018. Summary for Policymakers.
Timur Gül, Laura Cozzi, and Petr Havlik, "What Does Net-Zero Emissions by 2050 Mean for Bioenergy and Land Use?," International Energy Agency, May 31, 2021
"Pathway to Critical and Formidable Goal of Net-Zero Emissions by 2050 Is Narrow but Brings Huge Benefits, According to IEA Special Report," International Energy Agency (IEA), May 18, 2021.

ACTION 66

Third National Climate Assessment, GlobalChange.gov.
Jean-Francois Bastin et al., "The Global Tree Restoration Potential," *Science* 365, no. 6448 (July 2019): 76–79.
IPCC, 2019. Summary for Policymakers.

ACTION 67

Aditi Sen and Nafkote Dabi, Tightening the Net: Net Zero Climate Targets—Implications for Land and Food Equity, Oxfam, August 2021.
IPCC, 2019. Summary for Policymakers.

ACTION 68

Bronson W. Griscom, et al., "Natural Climate Solutions," PNAS 114, no. 44 (October 2017): 11645–11650.
T.A. Ontl and L.A. Schulte, (2012) Soil Carbon Storage. *Nature Education Knowledge* 3(10):35
"Soil Carbon Sequestration Impacts on Global Climate Change and Food Security," *Science*, June 11, 2004.
Playbook for Climate Action: Pathways for Countries and Businesses to Help Address Climate Change Today, The Nature Conservancy.
Judith D. Schwartz, "Soil as Carbon Storehouse: New Weapon in Climate Fight?," Yale Environment 360, March 4, 2014.

ACTION 69

Playbook for Climate Action: Pathways for Countries and Businesses to Help Address Climate Change Today, The Nature Conservancy.
IPCC, "Summary for Policymakers," in Special Report on the Ocean and Cryosphere in a Changing Climate, 2019.
Daniel A. Marchio et al., "Carbon Sequestration and Sedimentation in Mangrove Swamps Influenced by Hydrogeomorphic Conditions and Urbanization in Southwest Florida," *Forests* 7, no. 6 (May 2016): 116.
Kevin D. Kroeger et al., "Restoring Tides to Reduce Methane Emissions in Impounded Wetlands: A New and Potent Blue Carbon Climate Change Intervention," *Scientific Reports* 7: 11914, (September 2017).

ACTION 70

IPBES, Global Assessment Report on Biodiversity and Ecosystem Services of the Intergovernmental Science-Policy Platform on Biodiversity and Ecosystem Services, 2019.
IPBES, Summary for Policymakers of the Global Assessment Report on Biodiversity and Ecosystem Services of the Intergovernmental Science-Policy Platform on Biodiversity and Ecosystem Services, 2019.
"Why 30%," Campaign for Nature.
"Philanthropies pledge $5 billion to 'Protecting Our Planet Challenge,'" *Philanthropy News Digest*, September 22, 2021.
"Tropical Forest Restoration," Project Drawdown.
Bronson W. Griscom, et al., "Natural Climate Solutions."
Playbook for Climate Action: Pathways for Countries and Businesses to Help Address Climate Change Today, The Nature Conservancy.

Justine E. Hausheer, "Migration in Motion: Visualizing Species Movements Due to Climate Change," Cool Green Science, August 19, 2016.
Caitlin E. Littlefield et al., "Connectivity for Species on the Move: Supporting Climate-Driven Range Shifts," *Frontiers in Ecology and the Environment* 17, no. 5: 270–278, (June 2019).

ACTION 71

US Environmental Protection Agency "Benefits of Green Infrastructure," last updated September 7, 2022.
The Value of Green Infrastructure: A Guide to Recognizing Its Economic, Environmental and Social Benefits, Center for Neighborhood Technology, 2010.
American Rivers, the Water Environment Federation, the American Society of Landscape Architects, and ECONorthwest, Banking on Green: A Look at How Green Infrastructure Can Save Municipalities Money and Provide Economic Benefits Community-Wide, American Society of Landscape Architects, April 2012.
"Chapter 2: Our Changing Climate," Fourth National Climate Assessment, 2018.

ACTION 72

IPCC, "Summary for Policy Makers," in Climate Change 2021: The Physical Science Basis.
Mark Buchanan, "The Problem of Attribution," *Nature Physics* 17, no. 978 (2021).
"NYC CoolRoofs," NYC Business.
Dow Constantine, "Results of Heat Mapping Project Show Inequitable Impact of Hotter Summers, Will Inform Actions by King County and City of Seattle," King County, June 23, 2021.
US Environmental Protection Agency, "Using Trees and Vegetation to Reduce Heat Islands," last updated July 15, 2021.
Greg McPherson, James R. Simpson, Paula J. Peper, Scott E. Maco, and Quingfu Xiao "Municipal Forest Benefits and Costs in Five US Cities," *Journal of Forestry*, 103(8):411–416.
Nadja Popovich and Christopher Flavelle, "Summer in the City Is Hot, but Some Neighborhoods Suffer More," *New York Times*, August 9, 2019.
Brad Plumer and Nadja Popovich. "How Decades of Racist Housing Policy Left Neighborhoods Sweltering," *New York Times*, August 24, 2020.
Ladd Keith et al., "Deploy Heat Officers, Policies and Metrics," *Nature*, October 5, 2021.

ACTION 73

Mark Buchanan, "The Problem of Attribution."
IPCC, 2021. Summary for Policymakers.
"Western North American Extreme Heat Virtually Impossible Without Human-Caused Climate Change," World Weather Attribution, July 7, 2021.
Deepa Shivaram, "Heat Wave Killed an Estimated 1 Billion Sea Creatures, and Scientists Fear Even Worse," *NPR*, July 9, 2021.
Josie Fischels, "Photos: The Record-Breaking Heat Wave That's Scorching The Pacific Northwest," *NPR*, June 29, 2021.
US Environmental Protection Agency, "Climate Change Indicators: Heat-Related Deaths," last updated August 2, 2022.
The Lancet Countdown on Health and Climate Change, Policy Brief for the United States of America.
Jenessa Duncombe, "Climate Change Will Make Us Sicker and Lose Work Hours," *Eos*, November 18, 2019.
Drew Shindell et al., "The Effects of Heat Exposure on Human Mortality Throughout the United States," *GeoHealth* 3, (2020).
Ladd Keith et al., "Deploy Heat Officers, Policies and Metrics," *Nature*, October 5, 2021.
US Environmental Protection Agency, "Excessive Heat Events Guidebook in Brief," 430-B-06-006, June 2006.

ACTION 74

Ernani F. Choma et al., "Health Benefits of Decreases in On-Road Transportation Emissions in the United States from 2008 to 2017," Proceedings of the National Academy of Sciences 118, no. 51: e2107402118, (December 2021).
US Environmental Protection Agency, "Study Finds Exposure to Air Pollution Higher for People of Color Regardless of Region or Income," September 20, 2021.
Max Kozlov, "How Record Wildfires Are Harming Human Health," *Nature*, November 24, 2021.
Drew Costley, "Vehicle Emission Declines Decreased Deaths, Study Finds," *ABC News*, December 13, 2021.
US Environmental Protection Agency, Office of Transportation and Air Quality, Fast Facts: U.S. Transportation Sector Greenhouse Gas Emissions 1990–2018, June 2020.
US Environmental Protection Agency, "Criteria Air Pollutants," last updated August 9, 2022.
Climate Matters, "Pollen & Allergy Season," Climate Central, March 24, 2021.
Anjum Hajat, Charlene Hsia, and Marie S. O'Neill, "Socioeconomic Disparities and Air Pollution Exposure: A Global Review," *Current Environmental Health Reports* 2, no. 4 (December 2015): 440–450.
Christopher W. Tessum et al., "PM2.5 Polluters Disproportionately and Systemically Affect People of Color in the United States," *Science Advances* 7, no. 18, (April 2021).

ACTION 75

Centers for Disease Control and Prevention, "Illnesses on the Rise from Mosquito, Tick, and Flea Bites," last reviewed November 28, 2018.
Moritz U. G. Kraemer et al., "Past and Future Spread of the Arbovirus Vectors Aedes aegypti and Aedes albopictus," *Nature Microbiology* 4: 854–863, (2019).
US Environmental Protection Agency, "Climate Change Indicators: Lyme Disease," last updated August 2, 2022.
Climate Matters, "Hiking Hazards: Ticks and Poison Ivy," Climate Central, August 5, 2020.
"Climate Change Will Expose Half of World's Population to Disease-Spreading Mosquitoes By 2050," Yale Environment 360, March 5, 2019.
World Health Organization, "Climate Change and Health," October 30, 2021.
"Aedes aegypti—Factsheet for Experts," European Centre for Disease Prevention and Control (ECDC), last updated December 20, 2016.
Climate Matters, "This News Bites: More Mosquito Days," Climate Central, July 29, 2020.
National Institute for Occupational Safety and Health (NIOSH), "NIOSH Fast Facts: Protecting Yourself from Ticks and Mosquitoes," Centers for Disease Control and Prevention (CDC), DHHS (NIOSH) publication number 2010–119, April 2010.
"Prevent Tick and Mosquito Bites," Division of Vector-Borne Diseases, NCEZID, CDC, last reviewed February 9, 2022.

ACTION 76

"Yale Climate Opinion Maps 2021," Yale Program on Climate Change Communication (blog).

Peter D. Howe et al., "Geographic Variation in Opinions on Climate Change at State and Local Scales in the USA," *Nature Climate Change* 5, (June 1, 2015): 596–603.

Psychiatry.Org, "How Extreme Weather Events Affect Mental Health."

Tosin Thompson, "Young People's Climate Anxiety Revealed in Landmark Survey."

S. Clayton, C. M. Manning, K. Krygsman, and M. Speiser. (2017). *Mental Health and Our Changing Climate: Impacts, Implications, and Guidance.* Washington, D.C.: American Psychological Association, and ecoAmerica.

Paolo Cianconi, Sophia Betrò, and Luigi Janiri, "The Impact of Climate Change on Mental Health: A Systematic Descriptive Review," *Frontiers in Psychiatry* 11 (2020).

Tosin Thompson. "Young People's Climate Anxiety Revealed in Landmark Survey."

Glenn Albrecht et al., "Solastalgia: The Distress Caused by Environmental Change," Australasian Psychiatry: Bulletin of Royal Australian and New Zealand College of Psychiatrists 15 Suppl 1 (2007): S95-98.

Glenn Albrecht, "The Age of Solastalgia," *The Conversation*, August 7, 2012.

Brittany Harker Martin, "Brain Research Shows the Arts Promote Mental Health," *The Conversation*, June 9, 2020.

Psychiatry.Org, "Climate Change and Mental Health Connections."

ACTION 77

Julia Bentz, "Learning about Climate Change in, with and through Art," Climatic Change 162, (October 1, 2020): 1595–1612.

Liselotte J. Roosen, Christian A. Klöckner & Janet K. Swim (2018) Visual Art as a Way to Communicate Climate Change: A Psychological Perspective on Climate Change–Related Art, World Art, 8:1, 5-110.

Laura Kim Sommer and Christian Andreas Klöckner, "Does Activist Art Have the Capacity to Raise Awareness in Audiences?—A Study on Climate Change Art at the ArtCOP21 Event in Paris," *Psychology of Aesthetics, Creativity, and the Arts* 15, no. 1 (2021): 60–75.

Sarah Cascone, "Can Art Change Minds About Climate Change? New Research Says It Can—But Only If It's a Very Specific Kind of Art | Artnet News."

Brittany Harker Martin, "Brain Research Shows the Arts Promote Mental Health," *The Conversation*.

ACTION 78

Matthew M. Coggon et al., "Diurnal Variability and Emission Pattern of Decamethylcyclopentasiloxane (D5) from the Application of Personal Care Products in Two North American Cities," *Environmental Science & Technology* 52, no. 10 (May 15, 2018): 5610–18.

Kendra Pierre-Louis and Hiroko Tabuchi, "Want Cleaner Air? Try Using Less Deodorant," *New York Times*, February 16, 2018.

"Sustainable Beauty: The Beauty Industry's Role in Mitigating Climate Change," Beauty Packaging.

Patrizia Cinelli et al., 2019. "Cosmetic Packaging to Save the Environment: Future Perspectives," *Cosmetics* 6, no. 2: 26

Jea Morris, "Global Warming Is Coming for Our Beauty Bags," *ELLE*, November 22, 2021.

Patricia Megale Coelho et al., 2020. "Sustainability of Reusable Packaging–Current Situation and Trends," *Resources, Conservation & Recycling:* X 6.

ACTION 79

Nick Obradovich and James H. Fowler, "Climate Change May Alter Human Physical Activity Patterns," *Nature Human Behaviour* 1, no. 5 (April 24, 2017): 1–7.

"Climate and Sport," Nike.

Andrew Grundstein et al., "Exceedance of Wet Bulb Globe Temperature Safety Thresholds in Sports under a Warming Climate," Climate Research 58, no. 2: 183–91, (2013).

Paquito Bernard et al., "Climate Change, Physical Activity and Sport: A Systematic Review," *Sports Medicine* 51, no. 5: 1041–59, (2021).

ACTION 80

Ted Barrett, "Inhofe Brings Snowball on Senate Floor as Evidence Globe Is Not Warming, CNN Politics."

Minnesota Department of Natural Resources, "Climate Trends."

"On Thin Ice: How Climate Change Is Shaping Winter Recreation" Climate Central.

IPCC, 2019. Special Report on the Ocean and Cryosphere in a Changing Climate.

Protect Our Winters 2018 Annual Report.

Christine Chung, "After Warm Start to Snow Season, Colorado Resorts Look for Relief," *New York Times*, December 8, 2021.

Lauren Jackson, "Here's How Climate Change and Covid Are Transforming Skiing," *New York Times*, January 7, 2022.

"About POW," Protect Our Winters blog.

Torsten Masson and Immo Fritsche, "We Need Climate Change Mitigation and Climate Change Mitigation Needs the 'We': A State-of-the-Art Review of Social Identity Effects Motivating Climate Change Action," *Current Opinion in Behavioral Sciences*, Human Response to Climate Change: From Neurons to Collective Action, 42 (December 1, 2021): 89–96.

ACTION 81

US Department of the Interior, "New 5-Year Report Shows 101.6 Million Americans Participated in Hunting, Fishing & Wildlife Activities," September 7, 2017.

US Department of the Interior, US Fish and Wildlife Service, and US Department of Commerce, US Census Bureau. 2016 National Survey of Fishing, Hunting, and Wildlife-Associated Recreation.

US Bureau of Economic Analysis, "Outdoor Recreation."

D. Lipton et al., Ecosystems, Ecosystem Services, and Biodiversity. In *Impacts, Risks, and Adaptation in the United States: Fourth National Climate Assessment, Volume II* [Reidmiller, D.R., C.W. Avery, D.R. Easterling, K.E. Kunkel, K.L.M. Lewis, T.K. Maycock, and B.C. Stewart (eds.)]. U.S. Global Change Research Program, Washington, DC, USA, pp. 268–321, (2018).

Steven J. Dundas and Roger H. von Haefen, "The Effects of Weather on Recreational Fishing Demand and Adaptation: Implications for a Changing Climate," Journal of the Association of Environmental and Resource Economists 7, no. 2 (March 1, 2020): 209–42.

Sarah R. Weiskopf, Olivia E. Ledee, and Laura M. Thompson, "Climate Change Effects on Deer and Moose in the Midwest," *The Journal of Wildlife Management* 83, no. 4 (May 1, 2019): 769–81.

Valerio Barbarossa et al., "Threats of Global Warming to the World's Freshwater Fishes," *Nature Communications* 12, no. 1: 1701, (March 15, 2021).
"Beyond Season's End: A Path Forward for Fish and Wildlife in the Era of Climate Change, Adaptation Clearinghouse."
US Fish and Wildlife Service, Https://www.fws.gov/wsfrprograms/subpages/licenseinfo/Natl%20Hunting%20License%20Report%202021.pdf
"Gross Cost of Fishing Licenses in the United States from 2000 to 2021," Statista.
"Dingell-Johnson Act," Washington Department of Fish & Wildlife.
US Department of the Interior, "Secretary Zinke Announces More Than $1.1 Billion for Sportsmen & Conservation," March 20, 2018.

ACTION 82

Jennifer L. Lawless, "Becoming a Candidate: Political Ambition and the Decision to Run for Office" (Cambridge: Cambridge University Press, 2011).
US Census Bureau, "2017 Census of Governments–Organization."
Zoltan L. Hajnal ,"Why Does No One Vote in Local Elections?" *New York Times*, October 22, 2018.
"Who Votes For Mayor?", A Project of Portland State University and the Knight Foundation.

ACTION 83

Leiserowitz et al., *Politics & Global Warming*, (2021).

ACTION 84

Samuel A Markolf et al., "Pledges and Progress: Steps toward Greenhouse Gas Emissions Reductions in the 100 Largest Cities across the United States," Brookings, n.d., 34, (October 2020).
"State Climate Policy Maps," Center for Climate and Energy Solutions.
King County, "2020 Strategic Climate Action Plan."

ACTION 85

"View Environment Fundraisers on Gofundme, The World's #1 Most Trusted Fundraising Platform."

ACTION 86

Greta Thunberg Mocks World Leaders in "Blah, Blah, Blah" Speech, *BBC News*, 2021.
Heejin Han and Sang Wuk Ahn, "Youth Mobilization to Stop Global Climate Change: Narratives and Impact," *Sustainability* 12, no. 10 (May 2020): 4127.
"Secretary-General Launches Youth Advisory Group, Calling for Swift Action to Tackle Climate Change, Shape COVID-19 Recovery, Confront Injustice UN Meetings Coverage and Press Releases.
Wim Thiery et al., "Intergenerational Inequities in Exposure to Climate Extremes," *Science* 374, no. 6564: 158–60m (October 8, 2021).
Karen O'Brien, Elin Selboe, and Bronwyn M. Hayward, "Exploring Youth Activism on Climate Change: Dutiful, Disruptive, and Dangerous Dissent," *Ecology and Society* 23, no. 3 (October 2018).

ACTION 87

Kate Sambrook et al., "The Role of Personal Experience and Prior Beliefs in Shaping Climate Change Perceptions: A Narrative Review," *Frontiers in Psychology* 12, (2021).
Rachel I. McDonald, Hui Yi Chai, and Ben R. Newell, "Personal Experience and the 'Psychological Distance' of Climate Change: An Integrative Review," *Journal of Environmental Psychology* 44 (December 1, 2015): 109–18.
Christina Demski et al., "Experience of Extreme Weather Affects Climate Change Mitigation and Adaptation Responses," *Climatic Change* 140: 149–64, (January 1, 2017).

ACTION 88

Juliette N. Rooney-Varga et al., "Building Consensus for Ambitious Climate Action Through the World Climate Simulation," *Earth's Future* 9, no. 12, (December 2021).
"The En-ROADS Climate Workshop," Climate Interactive.
Cian Maher, "A New Wave of Indies Are Using Games to Explore Climate Change," *The Verge*, February 13, 2020.
"Play Games," NASA Climate Kids.
Andrew Perrin, "5 Facts about Americans and Video Games," Pew Research Center (blog).

ACTION 89

Thiery et al., "Intergenerational Inequities in Exposure to Climate Extremes," (2021).
IPCC, 2021. Summary for Policymakers.

ACTION 90

"K—12 Education: School Districts Frequently Identified Multiple Building Systems Needing Updates or Replacement," US Government Accountability Office (GAO) Report to Congressional Addressees, June 2020.
"Building for the Future: Investing in Climate Change Mitigation and Adaptation in Schools," K12 Climate Action.
"Schools," ASCE's 2021 Infrastructure Report Card blog, January 17, 2017.
Schools, Infrastructure Report Card, 2021.
"Opinion: Schools Can Help Us Build Back Better and Address Climate Change," The Hechinger Report, April 1, 2021.
"Building Retrofitting," Project Drawdown.

ACTION 91

Anya Kamenetz, "Most Teachers Don't Teach Climate Change; 4 In 5 Parents Wish They Did," *NPR*, April 22, 2019.
NPR, 2017.
Eric Plutzer et al., "Climate Confusion among U.S. Teachers," *Science* 351, no. 6274: 664–65, (February 12, 2016).
Leiserowitz et al., *Politics & Global Warming*, (2021).

ACTION 92

Anya Kamenetz, "Most Teachers Don't Teach Climate Change; 4 In 5 Parents Wish They Did," *NPR*, April 22, 2019.
Lydia Denworth, "Children Change Their Parents' Minds about Climate Change," *Scientific American*, May 16, 2019.
Danielle F. Lawson et al., "Children Can Foster Climate Change Concern among Their Parents," *Nature Climate Change* 9, 458–62, (June 2019).
Anya Kamenetz, "Climate Change Is Here. These 6 Tips Can Help You Talk to Kids about It," *NPR*, April 22, 2022.

ACTION 93

Stephan Lewandowsky et. al, *The Debunking Handbook 2020*, (2020).

John Cook, "Understanding and Countering Misinformation About Climate Change." In Handbook of Research on Deception, Fake News, and Misinformation Online. edited by Chiluwa, Innocent E., and Sergei A. Samoilenko, 281-306. Hershey, PA: IGI Global, 2019.
"Calling Bullshit," CallingBullShit.org.

ACTION 94

"GlobalChange.Gov," GlobalChange.gov.
"Fifth National Climate Assessment," GlobalChange.gov. (Full transparency, the author of this book is one of the report's authors.)

ACTION 97

Toby Bolsen, Risa Palm, Justin T. Kingsland, "The Impact of Message Source on the Effectiveness of Communications About Climate Change 2019."
Lauren Feldman and P. Sol Hart (2018), Is There Any Hope? How Climate Change News Imagery and Text Influence Audience Emotions and Support for Climate Mitigation Policies. *Risk Analysis*, 38: 585-602.
Matthew H. Goldberg et al., "Discussing Global Warming Leads to Greater Acceptance of Climate Science," *Proceedings of the National Academy of Sciences* 116, no. 30: 14804–5, (July 23, 2019).
Matto Mildenberger and Dustin Tingley, "Beliefs about Climate Beliefs: The Importance of Second-Order Opinions for Climate Politics," *British Journal of Political Science* 49, no. 4 (October 2019): 1279–1307.

ACTION 98

"#InDenial—Facebook's Growing Friendship With Climate Misinformation," Stop Funding Heat blog, November 2021.
Veronica Penney, "How Facebook Handles Climate Disinformation," *New York Times*, July 14, 2020.
Kari Paul, "Climate Misinformation on Facebook 'Increasing Substantially', Study Says," *The Guardian*, November 4, 2021.
"Climate Science Center," Facebook.
Lewandowsky et al., *The Debunking Handbook 2020.*

ACTION 99

"Environment & Sustainability," Social Impact at Penguin Random House blog.
Donnachadh McCarthy, "The Book Industry Fells Billions of Trees—It's Down to Authors to Demand to Be Printed on Recycled Paper," *The Independent*, February 12, 2021.
S. Subak and A. Craighill, "The Contribution of the Paper Cycle to Global Warming," *Mitigation and Adaptation Strategies for Global Change* 4, (June 1, 1999): 113–36.
Christine Ro, "Reducing the Environmental Toll of Paper in the Publishing Industry," Book Riot blog, February 11, 2021.
Matthew Schneider-Mayerson et al., "Environmental Literature as Persuasion: An Experimental Test of the Effects of Reading Climate Fiction," Environmental Communication 0: 1–16, (September 15, 2020).
"New Support for SDG13: Publishing-Related Organizations Commit to Climate Action," Publishing Perspectives, September 30, 2021.
Michelle Faverio and Andrew Perrin, "Three-in-Ten Americans Now Read e-Books," Pew Research Center blog.
"Holiday Gifts Which Consumers Plan to Buy United States 2021," *Statista*, November 8, 2021.

ACTION 100

Monica Y. Bartlett and David DeSteno, "Gratitude and Prosocial Behavior: Helping When It Costs You," *Psychological Science* 17, no. 4 (April 1, 2006): 319–25.

ILLUSTRATION NOTES

Infographics were derived from the author's text, except where noted below.

OVERVIEW

Page XVI: Hannah Ritchie and Max Roser, "Emissions by Sector," Our World in Data, 2020.
Page XIX: "Which Countries Have a Net Zero Carbon Goal?," Climate Home News, June 14, 2019.
Pages XXII–XXIII: NCA4 Graphic from Track the State of the Science entry.

ACTION 2

Euan Davis and Eduardo Plastino, "What It Takes To Be a Future-Fit Business Today," *Forbes*, May 3, 2022.
Ritchie and Roser, "Emissions by Sector," Our World in Data.

ACTION 3

Seth Wynes and Kimberly A. Nicholas, "The Climate Mitigation Gap: Education and Government Recommendations Miss the Most Effective Individual Actions," Environmental Research Letters 12, no. 7 (July 2017).

ACTION 4

A. Leiserowitz et al., *Politics & Global Warming*, September 2021. Yale University and George Mason University. New Haven, CT: Yale Program on Climate Change Communication.

ACTION 5

Geoffry Supran and Naomi Oreskes, "The Forgotten Oil Ads That Told Us Climate Change Was Nothing," *The Guardian*, November 18, 2021.

ACTION 6

Hannah Ritchie, "What Are the Safest and Cleanest Sources of Energy?" Our World in Data, 2020.

ACTION 7

Hannah Ritchie and Max Roser, "Energy Mix," Our World in Data, 2020.

ACTION 9

Jocelyn Timperley, "Explainer: These Six Metals Are Key to a Low-Carbon Future," CarbonBrief, April 12, 2018.

ACTION 12

Brad Plumer, Nadja Popovich, and Blacki Migliozzi, "Electric Cars Are Coming. How Long Until They Rule the Road?" *New York Times*, March 10, 2021.

ACTION 13

"Drive More Efficiently," Fueleconomy.gov.
"Drive Less and Drive Clean," CoolCalifornia.org.

ACTION 15

Stefan Gössling and Andreas Humpe, "The Global Scale, Distribution and Growth of Aviation: Implications for Climate Change," *Global Environment Change* 65, (November 2020).
"World Aviation and the World Economy," International Civil Aviation Organization.

ACTION 16

Jocelyn Timperley, "Should We Give Up Flying for the Sake of the Climate?," *BBC*, February 18, 2020.

ACTION 22

Rebecca Dzombak, "Remote Work May Be Keeping Some Cities' Air Cleaner," Phys.org, October 13, 2021.

Robert McSweeney and Ayesha Tandon, "Global Carbon Project: Coronavirus Causes 'Record Fall' in Fossil-Fuel Emissions in 2020," CarbonBrief, December 11, 2020.

Dawit Habtemariam, "Survey: Virtual to Replace One-Quarter of 2022 Travel Volume, on Average," *Business Travel News*, July 26, 2021.

Kim Parker, Juliana Menasce Horowitz, and Rachel Minkin, "How the Coronavirus Outbreak Has—and Hasn't—Changed the Way Americans Work," Pew Research Center, December 9, 2020.

"Telepresence," Project Drawdown.

ACTION 26

Manish Ram et al., "Job Creation During a Climate Compliant Global Energy Transition across the Power, Heat, Transport, and Desalination Sectors by 2050," *Energy* 238, Part A, (January 2022).

ACTION 28

Hannah Ritchie, "You Want to Reduce the Carbon Footprint of Your Food? Focus on What You Eat, not Whether Your Food Is Local," Our World in Data, 2020.

ACTION 29

"Reducing Food Waste," Congresswoman Chellie Pingree, 1st District of Maine.

ACTION 30

US Environmental Protection Agency, "Reducing the Impact of Wasted Food by Feeding the Soil and Composting," last updated February 12, 2022.

ACTION 31

Brent R. Heard, et al., "Comparison of Life Cycle Environmental Impacts from Meal Kits and Grocery Store Meals," *Resources, Conservation and Recycling* 147 (August 2019): 189–200.

ACTION 32

Robert Crawford, "Home-Delivered Food Has a Huge Climate Cost. So Which Cuisine Is the Worst Culprit?," *The Conversation*, January 18, 2021.

ACTION 33

Saloni Shah and Dan Balustein-Rejto, "Can Cultivated Meat Live Up to Its Environmental Promise?," *The Breakthrough Institute*, May 13, 2020.

ACTION 34

J. Poore and T. Nemecek, "Reducing Food's Environmental Impacts through Producers and Consumers," *Science* 360, no. 6392 (June 1, 2018): 987–92.

ACTION 36

Dave Reay, "Climate-Smart Chocolate," in *Climate-Smart Food* (Palgrave Pivot: Cham, Switzerland, 2019).

ACTION 37

Eric Asimov, "How Climate Change Impacts Wine," *New York Times*, October 14, 2019.

ACTION 38

Carmen Nab and Mark Maslin, "Life Cycle Assessment Synthesis of the Carbon Footprint of Arabica Coffee: Case Study of Brazil and Vietnam Conventional and Sustainable Coffee Production and Export to the United Kingdom," *Geo: Geography and Environment* 7, no. 2 (July–December 2020).

ACTION 39

Seafood Carbon Emissions Tool.

ACTION 40

Nabil Nasr et al., "Re-defining Value – The Manufacturing Revolution: Remanufacturing, Refurbishment, Repair and Direct Reuse in the Circular Economy." A Report of the International Resource Panel. United Nations Environment Programme, Nairobi, Kenya. pp 272.

ACTION 41

K. Niinimäki et al., The environmental price of fast fashion. Nat Rev Earth Environ 1, 189–200, (2020).

Andrew Morlet et al., "A New Textiles Economy: Redesigning Fashions' Future" Ellen Macarthur Foundation updated January 12, 2017.

Pauline Chrobot et al., "Measuring Fashion: Insights from the Environmental Impact of the Global Apparel and Footwear Industries" Quantis, pp 65, (2018).

Anna, Niklas, and Jesper, "List of Things That Weigh One Ton," Weight of Stuff.

ACTION 42

Kendyl Salcito, "Why Cloth Diapers Might Not Be the Greener Choice, After All," *Washington Post*, May 8, 2015.

ACTION 45

"Microplastics in Majority of Oregon Shellfish, Study Finds," *Seattle Times*, November 12, 2019.

Wikipedia, s.v. "Microplastics," last edited August 27, 2022.

ACTION 46

Graham Charlton, "Ecommerce Returns: 2020 Stats and Trends," SaleCycle, January 15, 2020, last revision by Brad Ward July 8, 2021.

ACTION 47

Miguel Jaller, Anmol Pahwa, and Seth Karten, "UC Davis Transportation and Climate Blog: Keeping e-Commerce Environmentally Friendly—What Consumers Can Do," UC Davis Institute of Transportation Studies Sustainable Freight, December 2, 2020.

ACTION 48

Sandra Wilson, "I'm a Bit of a Modern-Day Alchemist, Recovering Gold from Old Mobile Phones," *The Conversation*, May 21, 2020.

"E-Waste Monitors Report," United Nations Institute for Training and Research.

ACTION 53

Build a Kit, Ready.gov, last updated May 5, 2022.

A.C. Shilton, "Pack a 'Go Bag' Now," *New York Times*, September 19, 2020.

ACTION 54

Energy for Single-Family Homes: What You Need to Know, Clean Energy Resource Teams, 2021.

Catherine Evans, "Climate Change: How Can You Make Your Home Eco-Friendly?," *BBC*, June 20, 2021.

"Savings," Expert Home Improvements, EnergyStar.

ACTION 56
Chiara Delmastro, Cooling, International Energy Agency (IEA), November 2021.
The Future of Cooling, International Energy Agency (IEA), May 2018.

ACTION 57
Yasmina Abdelilah et al., Renewables 2021: Analysis and Forecast to 2026, International Energy Agency (IEA), December 2021.

ACTION 61
Simone Conradi, "Cherry Tree and Global Warming: To Bloom or Not to Bloom?," *Towards Data Science*, April 13, 2017.

ACTION 62
US Environmental Protection Agency, National Overview: Facts and Figures on Materials, Wastes and Recycling, last updated July 31, 2022.

ACTION 66
Hannah Ritchie, "The World Has Lost One-Third of Its Forest, but an End of Deforestation is Possible," Our World in Data, February 9, 2021.
Gabrielle Lipton, "4.06 Billion Remaining Hectares, and Other New Numbers on Forests . . . But What Do They Mean?" *Landscape News*, May 14, 2020.

ACTION 68
"Soil Composition," University of Hawai'i at Mānoa, Soil Nutrient Management for Maui County, College of Tropical Agriculture and Human Resources (CTAHR).

ACTION 70
"A Transboundary Adaptation Plan for Our Shared Natural Systems," Blueprint for a Resilient Cascadia, Cascadia Partner Forum.

ACTION 71
Melissa Denchak, "Green Infrastructure: How to Manage Water in a Sustainable Way," NRDC, July 25, 2022.

ACTION 73
"Climate Action Benefits: Extreme Temperature," US Environmental Protection Agency, 2017.

ACTION 75
Climate Matters, "Hiking Hazards: Ticks and Poison Ivy," Climate Central, August 5, 2020.
Climate Matters, "This News Bites: More Mosquito Days," Climate Central, July 29, 2020.

ACTION 77
Warming Stripes illustration, Ed Hawkins, University of Reading, 2016.

ACTION 79
"Move to Zero: Tennis" Nike.

ACTION 80
"On Thin Ice: How Climate Change Is Shaping Winter Recreation," Climate Central, January 23, 2020.

ACTION 81
"License Dollars at Work," Minnesota Department of Natural Resources.

ACTION 82
"How Many Politicians Are There in the USA?" PoliEngine.

ACTION 84
Center for Climate and Energy Solutions. "U.S. State Climate Action Plans" Last updated August 2022.

ACTION 86
Portraits based on photos by: Etown.org (Martinez); Extinction Rebellion UK YouTube video (Thunberg) Angela DeCastro (Tan); Clement Guerra (Gualinga); TED Conferences (Nyathi); Joel Carrett/EPA-EFE/Shutterstock (Raj-Seppings).

ACTION 89
Gavin, "30 Years after Hansen's Testimony," RealClimate, Jun 21, 2018.

ACTION 91
Leiserowitz et al., *Politics & Global Warming*, (2021).

ACTION 96
Portraits based on photos by: The Alaska Center (Quintyne); Wednesday Journal of Oak Park and River Forest (Cuneen); Media Impact Funders (Kumari Drapkin); MaineSeniorCollege.org (Blackman); Center for New Democratic Processes (Goodnough); NM Healthy Soil Working Group (Skeets); Annie Mulligan/*Houston Chronicle* (Bullard); Midnight's Farm (Bill and Van De Putte); Jamie Stroble (Stroble).

ACTION 97
Jennifer Marlon et al., Yale Climate Opinion Maps 2021. Yale Program on Climate Change Communication. Last updated February 23, 2022.

ACTION 98
Stephan Lewandowsky et al., *The Debunking Handbook 2020*, 2020.

INDEX

Page numbers in *italic* refer to graphics.

D

E

F

G

H

I

L

M

N

O

P

R

S

ABOUT THE AUTHOR

PHOTO BY MARK D. STONE

DR. HEIDI ROOP is the Director of the University of Minnesota Climate Adaptation Partnership and an Assistant Professor of Climate Science and Extension Specialist at the University of Minnesota. Her research and Extension programs combine cutting-edge climate science and effective science communication to increase the use and integration of climate change information in decision-making at a range of scales—from city and state to national and international levels. Her climate science research takes her around the world from Antarctica to California to the shores of Lake Superior. She is also an affiliate assistant professor in the School of Public Health at the University of Washington, and serves as an expert advisor to organizations and agencies as they seek to build resilience to climate change.

ABOUT THE ILLUSTRATOR

PHOTO BY LAURA GOFF

JOSHUA M. POWELL is the author of *The Pacific Crest Trail: A Visual Compendium*. He works as a graphic designer and lives in the Inland Northwest with his wife and son.